Analysis Report On Firefighter Fatalities in the United States in 1994

Prepared for the

United States Fire Administration
Federal Emergency Management Agency
Contract No. EMW-95-C-4713

September 1995

ACKNOWLEDGMENTS

This study of firefighter fatalities would not have been possible without the cooperation and assistance of many members of the fire service across the United States. Members of individual fire departments, chief fire officers, the National Interagency Fire Center, U.S. Forest Service personnel, the U.S. military, the Department of Justice, and many others contributed important information for this report.

This analysis was conducted by TriData Corporation of Arlington, Virginia, for the United States Fire Administration under contract EMW-95-C-4713. Kathy Gerstner of the United States Fire Administration was the Project Officer. J. Gordon Routley of TriData served as Project Supervisor and Chief Analyst. Patricia Frazier and Philip Schaenman provided statistical analysis of NFIRS data, and Jeff Stern, Katherine Ivey, and Reade Bush conducted research and analysis.

We acknowledge that firefighting is a dangerous profession, and tragedies will occur from time to time. This is the risk all firefighters accept every time they respond to an emergency incident; however, the risk can be greatly reduced through efforts to increase firefighter safety. This report is dedicated to those firefighters who have made the ultimate sacrifice in 1994. May the lessons learned from their passing not go unheeded.

CONTENTS

BACKGROUND

For two decades, the United States Fire Administration (USFA) has kept track of firefighter fatalities and conducted an analysis of the fatalities that occur each year. Through the collection of information on the causes of firefighter deaths, the USFA is able to focus efforts on specific problems and direct efforts towards finding solutions to reduce the number of firefighter fatalities in the future. This information is also used to measure the effectiveness of current efforts directed toward firefighter health and safety.

The USFA also maintains a list of firefighter fatalities and their next of kin invited to the annual Fallen Firefighter Memorial Service, which is held at the National Fire Academy in Emmitsburg, Maryland every fall.

INTRODUCTION

The objective of this study was to identify on-duty firefighter fatalities, occurring in the U.S. in 1994, and analyze the circumstances surrounding firefighter fatalities. The study is intended to help identify potential approaches that could reduce the number of deaths that occur each year. In addition to the 1994 findings, this study includes a special analysis of wildland firefighting fatalities, which claimed an unusually high number of lives this year, and an analysis of risk management and recognition in the fatal incidents.

This report continues a series of annual studies by the U.S. Fire Administration of firefighter fatalities in the United States.

Who Is a Firefighter?

For the purpose of this study, the term *firefighter* *covers* all members of organized fire departments, including career and volunteer firefighters; full-time public safety officers acting as firefighters; state and federal government fire service personnel; including wildland firefighters; and privately employed firefighters, including employees of contract fire departments and trained members of industrial fire brigades, whether full or part-time. This also includes contract personnel working as firefighters or assigned to work in direct support of fire service organizations.

Under this definition, the study includes not only local and municipal firefighters, but also seasonal and full-time employees of the United States Forest Service, the Bureau of Land Management, the Bureau of Indian Affairs, the Bureau of Fish and Wildlife, the National Park Service, and state wildland agencies; prison inmates serving on firefighting crews; firefighters employed by other governmental agencies such as the United States Department of Energy; military personnel performing assigned fire suppression activities; and civilian firefighters working at military installations.

What Constitutes an On-duty Fatality?

The term *on-duty* refers to being involved in operations at the scene of an emergency, whether it is a fire or non-fire incident; being enroute to or from an incident; performing other officially assigned duties such as training, maintenance, public education, inspection, investigations, court testimony and fund-raising; and being on-call, under orders, or on stand-by duty, except at the individual's home or place of business.

On-duty fatalities include any injury or illness sustained while on-duty that proves fatal. These fatalities may occur on the fireground, in training, while responding to or returning from alarms, or while performing other duties that support fire service operations.

A common example of a fatal illness incurred on duty is a heart attack. Fatalities attributed to illnesses would also include a communicable disease contracted while on duty that proved fatal, where the disease could be attributed to an occupational exposure.

Accidents that claim the lives of on-duty firefighters are included in the analysis, whether or not they are directly related to emergency incidents. In 1994, this category includes a career firefighter who died when he fell off the roof of his fire station and a wildland fire management officer who was killed when a 106mm rifle exploded, while he was being trained for avalanche control duties.

Injuries and illnesses are also included in cases where death is considerably delayed after the original incident. When the incident and the

---

1A volunteer responding from home or work would be considered "on-duty" from the moment he is called to respond to an alarm, whether by pager, radio, house siren, or other means. He would remain "on-duty" until he had returned from the alarm.

death occur in different years, the analysis counts the fatality as having occurred in the year that the incident occurred. For example, a firefighter died in 1994 of medical complications that resulted from an injury that was incurred on duty in 1979. Because his death was the result of the 1979 injury, this case was counted as a 1979 fatality for statistical purposes, and is not included in the 104 fatalities for 1994 that were analyzed in this report. Since the death occurred in 1994, he will be included in the 1994 annual Fallen Firefighter Memorial Service at the National Fire Academy, and his name will be included on the list of firefighters who died in 1994.

There is no established mechanism for identifying fatalities that result from illnesses that develop over long periods of time, such as cancer, which may be related to occupational exposure to hazardous materials or products of combustion. It has proven to be very difficult over several years to fully evaluate occupational illness as a causal factor in firefighter deaths, because of the limitations in the ability to track the exposure of firefighters to toxic hazards and the potential long-term effects of such an exposure.

Sources of Initial Notification

As an integral part of the ongoing program to collect and analyze fire data, the U.S. Fire Administration solicits information on firefighter fatalities from the U.S. fire service and a wide range of other sources. These include the United States Fire Administration, and the Public Safety Officer's Benefit Program (PSOB) administered by the Department of Justice. Other sources include the Occupational Safety and Health Administration (OSHA), the U.S. military, the National Interagency Fire Center, and other federal agencies.

The Fire Administration also receives notification from fire service organizations such as the International Association of Fire Chiefs (IAFC), the International Association of Fire Fighters (IAFF), the National Fire Protection Association (NFPA), the National Volunteer Fire Council (NVFC), state fire marshals, state training organizations, other state and local

organizations, and fire service publications. The USFA also keeps track of
fatal fire incidents as part of the Major Fire Investigations Project and
maintains an ongoing analysis of data from the National Fire Incident
Reporting System (NFIRS) for the production of Fire in the United States.

Procedure for Including a Fatality in the Study

After notification of a fatal incident is received from any source, initial
contact is made by telephone with local authorities to verify the incident, its
location and jurisdiction, and the fire department or agency involved.
Further information may be obtained from the chief of the fire department or
his designee over the phone or by data collection forms, for both the deceased
firefighter and the fire incident. After the information is obtained by the
contractor, a determination is made as to whether the individual qualifies as
an on-duty firefighter fatality, according to the previously described criteria.
Additional information may be requested, either by follow-up with the fire
department directly, or through state vital records offices or other agencies.
The final determination as to whether a fatality qualifies as an on-duty death
for inclusion in the statistical analysis, and in the Fallen Firefighter
Memorial Service, is made by the United States Fire Administration.

Information that is routinely requested includes NFIRS-1 (incident)
and NFIRS-3 (fire service casualty) reports, the fire department's own
incident reports and internal investigation reports, copies of death
certificates or autopsy results, special investigative reports such as those
produced by the USFA or NFPA, police reports, photographs and diagrams,
and newspaper or media accounts of the incident. The same criteria have
been used for this study as in previous annual studies that were conducted by
the NFPA.

1994 FINDINGS

One-hundred four (104) firefighters died while on duty in 1994.[2] This is a significant rise after two consecutive years that produced the lowest number of firefighter fatalities since the USFA has kept records. The increase is primarily attributable to a significant increase in wildland firefighter fatalities, up from 8 deaths in 1993 to 38 deaths in 1994, including 14 firefighters killed in a single incident on Storm King Mountain in Colorado. Even with this exceptional incident, the total of 104 fatalities is still the third lowest number of fatalities recorded in the 18 years that this data has been reported, and continues the long term trend of reduced fatalities that began in 1979.

Deaths in 1994 increased approximately 35 percent from 1993. This increase of 27 deaths is shown in Figure 1. The number of fatalities involving wildland firefighters increased by 30 deaths from the previous year. The 104 deaths resulted from 81 incidents.

The fatalities included 38 volunteer firefighters; 34 career firefighters; 20 seasonal wildland firefighters; 4 career wildland personnel; 6 contract aircraft crew members; 1 industrial emergency response team member; and 1 civilian assigned to a military reserve unit. Ninety-seven of the fatalities were men; seven were women.

---

2 A total of 107 firefighter fatalities were reported in 1994. Three of these deaths were attributed to incidents that occurred in prior years. Two firefighters died in 1994 of complications from prior injuries, one from an accident in 1979 and one from a heart attack in 1991. Another firefighter died of AIDS which he is believed to have contracted while performing emergency medical services in the 1980s: since the exact date of exposure hasd not been documented, we have listed his death as a 1991 statistic, the year in which he was first diagnosed with HIV. Those deaths have been added to the statistics for the year in which the incident occurred, consistent with past USFA analysis. The data from two other incidents that occurred in 1994 is included in this analysis, although the firefighters actually died in 1995.

7

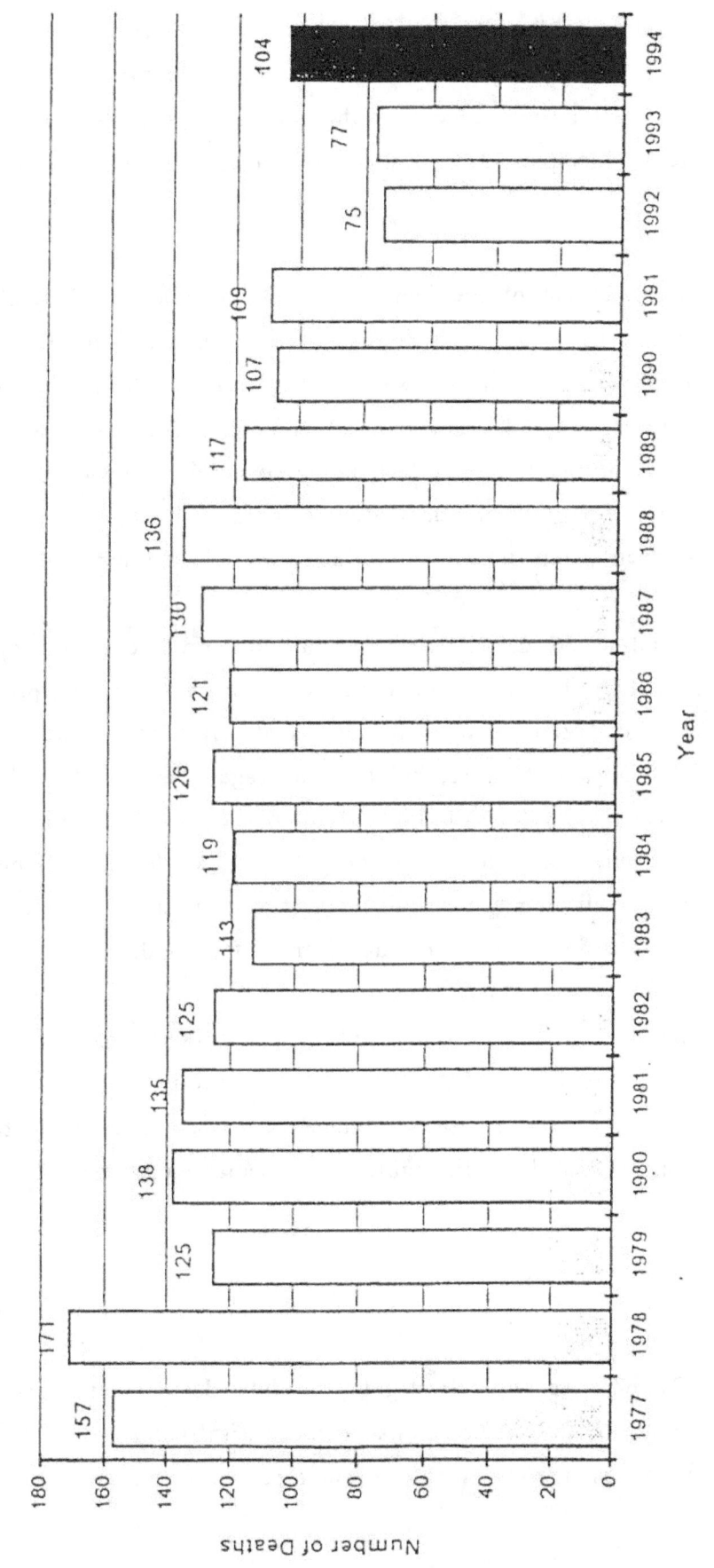

Figure 1

On-Duty Firefighter Deaths 1977-1994

In addition to the Storm King Mountain fire, there were five other incidents that caused more than one firefighter death: two airplane crashes (five deaths), one helicopter crash (three deaths), a high rise apartment building fire in Memphis (two deaths), and a church fire in Philadelphia (two deaths).

A large segment of the increase in total number of firefighter deaths in 1994 can be attributed to the high loss of lives in wildland fires. A total of 38 deaths were associated with wildland firefighting, including 22 that occurred on the fireground and 14 that involved personnel enroute to wildland fires. Two additional deaths involved full-time personnel who were employed in positions that were primarily dedicated to wildland fire protection, although their deaths were from injuries sustained while performing other duties.

The total of 36 deaths attributed to wildland fireground operations or response to wildland fires is considerably higher than in any previous year that has been analyzed. Fireground deaths at wildland incidents have varied considerably from year to year, but have averaged between 8 and 10 in the most recent years. The loss of 14 firefighters in the Storm King Mountain incident comprises a significant portion of the total deaths for the year; however, the overall increase remains significant, even if the number of lives lost in this single, incident is subtracted from the total.

The number of deaths associated with wildland fires in 1994 is a major cause for concern within the wildland fire community and has caused efforts to be initiated on several fronts to reduce the risks and increase the safety related training of wildland firefighters. A special analysis of this subject is included in this report.

Type of Duty

In 1994, 84 percent of firefighter on-duty deaths were associated with emergency incidents (Figure 2). This includes firefighters who died while responding to an emergency, while at the emergency scene, or after the

9

Figure 2

Firefighter Deaths While Performing Emergency Duty - 1994

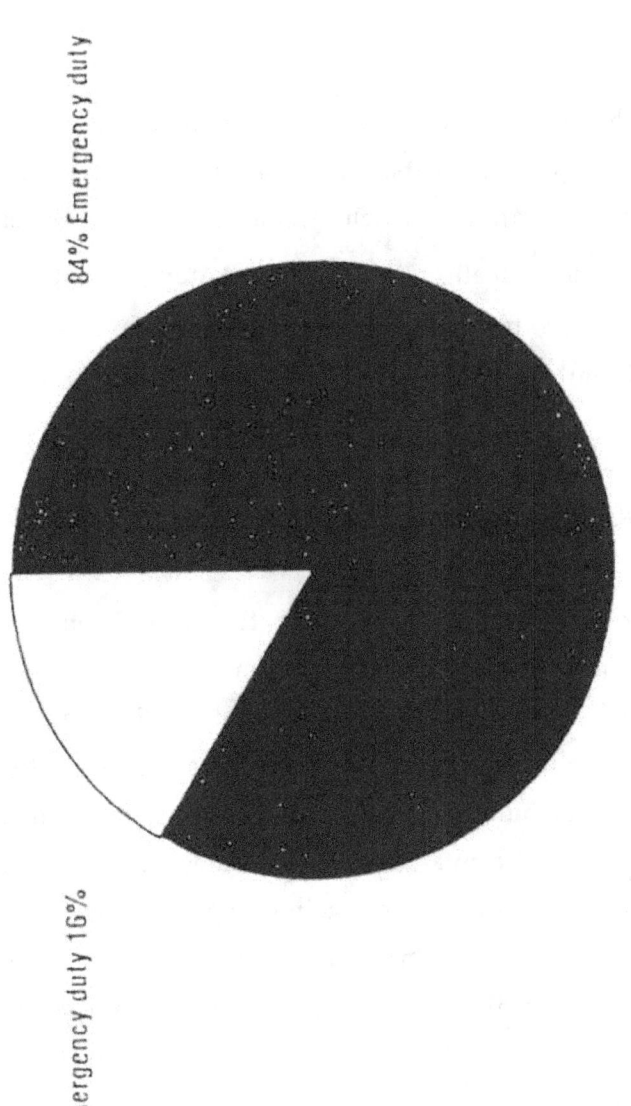

Non-emergency duty 16%

84% Emergency duty

emergency. Non-emergency activities accounted for 16 percent of the firefighter fatalities for 1994. Non-emergency duties included training, administrative activities, or performing other functions that are not related to an emergency incident. Two firefighters died in their sleep of massive heart attacks while on duty and are included in the non-emergency category. One volunteer firefighter suffered a fatal heart attack doing clerical work at the firehouse.

A distribution of deaths by type of duty being performed is shown in Figure 3. As in previous years, in 1994 the largest number of deaths resulted from fireground operations, which accounted for 57 percent of the fatalities. Of the 60 fireground deaths, 29 were attributed to smoke inhalation[3], 8 from burn injuries, and 3 from trauma (including two pilots killed when their plane crashed while dropping retardant on a wildfire). A total of 19 heart attack deaths were attributed to fireground operations, including one that occurred after returning from a fire. One firefighter was electrocuted and one died of an embolism after being injured in a fall at a fire.

The second largest category after fireground operations was responding to and returning from emergency incidents, which accounted for 22 deaths in 1994. This was also the second leading cause of deaths in 1993. Two firefighters and the patient they were transporting to the hospital were killed when their ambulance was struck head-on by a tractor-trailer truck while enroute to the hospital. Another 8 fatalities occurred in fire apparatus crashes, with rollovers being the leading type of fatal accident. Aircraft crashes while enroute to wildland fires accounted for 6 deaths. Five firefighters suffered fatal heart attacks while enroute or returning from alarms. Only one volunteer firefighter was killed in 1994 in a crash of a personal vehicle.

---

[3] All of the 14 firefighters who died at Storm King Mountain were listed as smoke inhalation deaths, the primary cause of death according to their death certificates.

11

Figure 3

Firefighter Deaths by Type of Duty - 1994

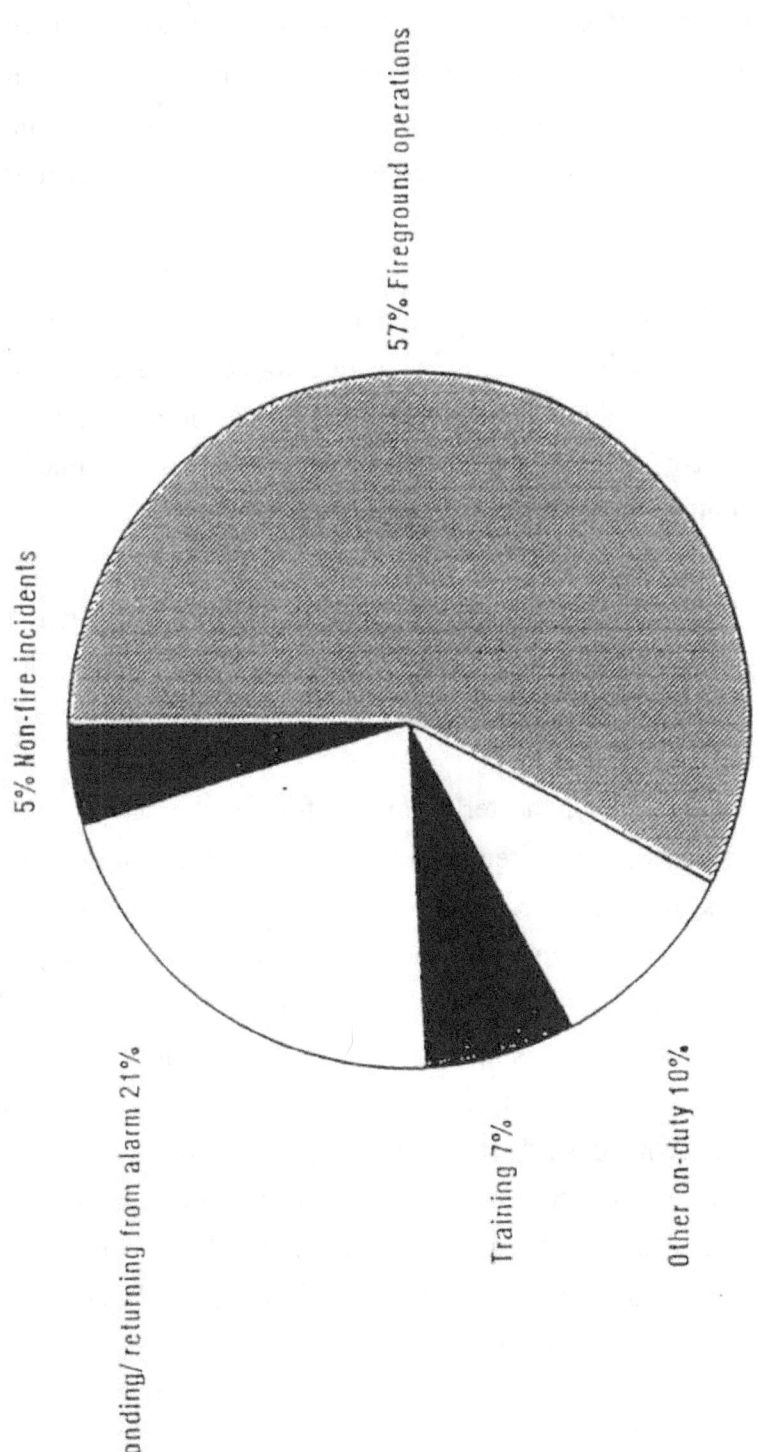

5% Non-fire incidents

57% Fireground operations

Responding/ returning from alarm 21%

Training 7%

Other on-duty 10%

Five deaths were related to activities at the scene of non-fire emergency incidents. Two of the victims were struck by vehicles: one firefighter was killed while on an emergency medical call, and a fire-police officer died while directing traffic at the scene of an incident. Three died of heart attacks: one firefighter died after extricating a patient from an automobile accident, another died after returning from an EMS incident in very hot weather, and a third died after helping to carry a patient to an ambulance.

Seven deaths occurred during training - none involving live fire training. Five of the training deaths were due to heart attacks and one firefighter died of a cerebral aneurysm. A fire management officer was killed when a 106mm recoilless rifle exploded while he was being trained to use it for avalanche control, part of his collateral duties.

Ten deaths occurred during other non-emergency duty activities.[4]

Cause and Nature of Fatal Injury or Illness

As used in this study, the term cause refers to the action, lack of action, or circumstances that directly resulted in the fatal injury, while the term *nature* refers to the medical nature of the fatal injury or illness, or what is often. referred to as the cause of death. Often, the fatal injury is the result of a chain of events, the first of which is recorded as the cause. For example, if a firefighter is struck by a collapsing wall, becomes trapped in the debris, runs out of air before being rescued, and died of asphyxiation, the cause of the fatal injury is recorded as "struck by collapsing wall" and the nature of the fatal injury is "asphyxiation." Likewise, if a wildland firefighter is overrun by a fire and dies of burns, the cause of the death would be listed as "caught/trapped," and the nature if death would be "burns." This follows the

---

4 A summary of all the firefighter fatality incidents is included as Appendix A.

convention used in NFIRS casualty reports, which are based on NFPA Fire Reporting standards.

Figure 4 shows the distribution of deaths by cause of fatal injury or illness. As in most previous years, the largest category is stress or overexertion, which was the listed as the primary factor in 35 percent of the deaths. The act of firefighting has been shown to be one of the most physically demanding activities that the human body performs, and deaths from stress are usually from heart attacks. Of the 37 stress related fatalities in 1994, 36 firefighters died of stress related heart attacks and one firefighter died of an aneurysm.

The second leading cause of firefighter fatalities was being caught or trapped, accounting for 36 firefighter fatalities (35%). Seventeen firefighters were overrun by rapidly moving brush or wildland fires. Nine firefighters were trapped by rapidly changing fire conditions inside burning structures and seven apparently became disoriented or lost and died in building fires. Two firefighters died as a result of becoming trapped in structural collapses. A fire chief died of burns after being caught in an explosion while directing operations outside a burning garage.

The third leading cause of firefighter fatalities was being struck by or coming in contact with an object. Of the 26 firefighters (25%) who died in these incidents, 11 were involved in vehicle accidents, 8 died in aircraft crashes, 2 were struck by vehicles while at the scene of an emergency, one was struck by a falling tree, one was struck by a helicopter rotor blade, one was hit by shrapnel when a 106mm recoilless rifle misfired and exploded, and one was struck by falling debris at a fire, and one was electrocuted when he came in contact with an electric line power line while carrying a chainsaw down an aerial ladder from the roof of a fire building.

Figure 4

Firefighter Deaths by Cause of Fatal Injury 1994

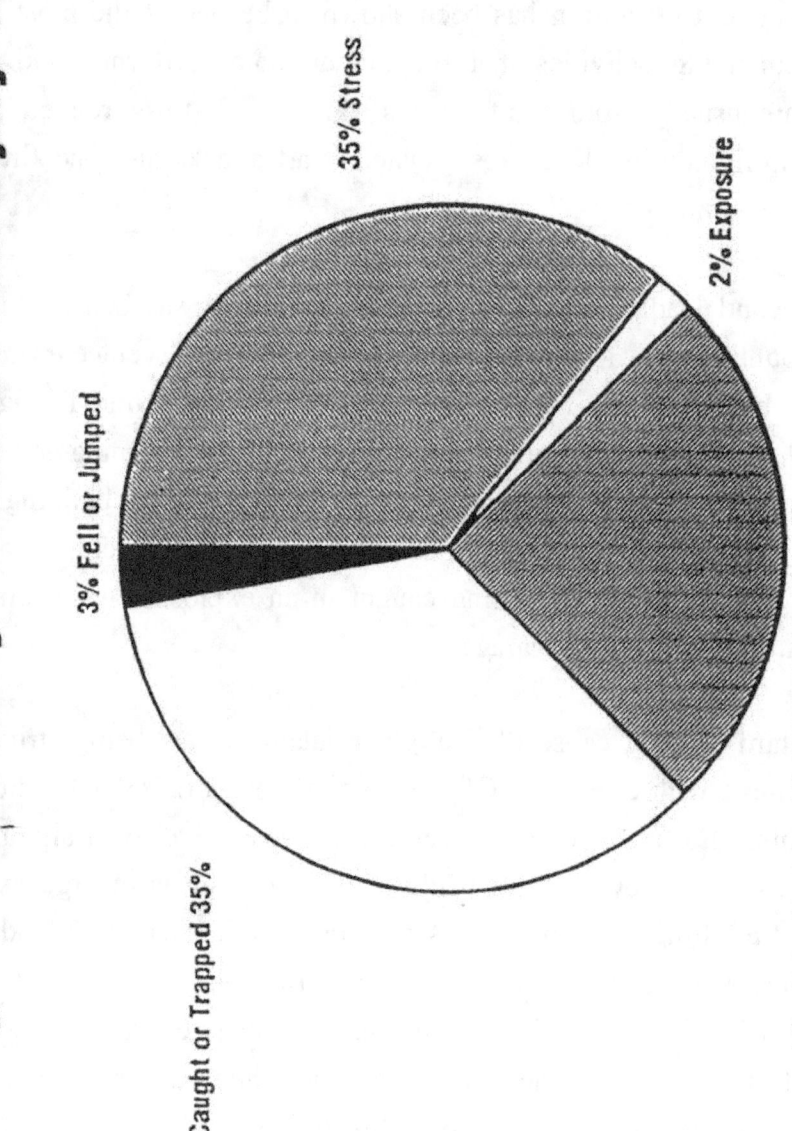

3% Fell or Jumped

35% Stress

2% Exposure

Caught or Trapped 35%

Struck by/ Contact with Object 25%

Three firefighters (3%) died as a result of falls. One firefighter died when he fell through a hole in the floor while checking a room for fire extension. A second firefighter, who had suffered a broken ankle when-he fell down a flight of stairs at a fire, died of an embolism that was caused by his injury. A third firefighter died when he fell off the roof of his fire station.

Exposure to smoke or toxic gases was listed as the causal factor in deaths (2%). A chief officer suffered a fatal heart attack after breathing toxic gases while performing overhaul at a house fire. He was not wearing any protective gear or SCBA. Another firefighter suffered a fatal heart attack after being overcome by heat and toxic gases while standing-by at a controlled burn.

Figure 5 shows the distribution of the 104 deaths by the medical nature of the fatal injury or illness. Thirty-nine (39) firefighters died of heart attacks in 1994: 36 caused by stress, 1 caused by an embolism, and 2 attributed to the inhalation of toxic gases on the fireground. At least 13 of the firefighters had known high risk factors for heart attacks, including prior heart conditions, high blood pressure, or obesity, including two who continued to perform firefighting activities after bypass surgery. Autopsy results[5] indicated that some coronary artery disease was present in most of the cases where medical records were available. Obesity and poor physical fitness were noted as factors in the deaths of several of the heart attack victims, including a 26-year old firefighter who was 5'6" tall and over 275 pounds when she died of a heart attack during a training exercise. At least one of the firefighters had a genetic heart defect that would not have been discovered during a routine physical.

---

5 Autopsy results and medical records were not available for all heart attack victims.

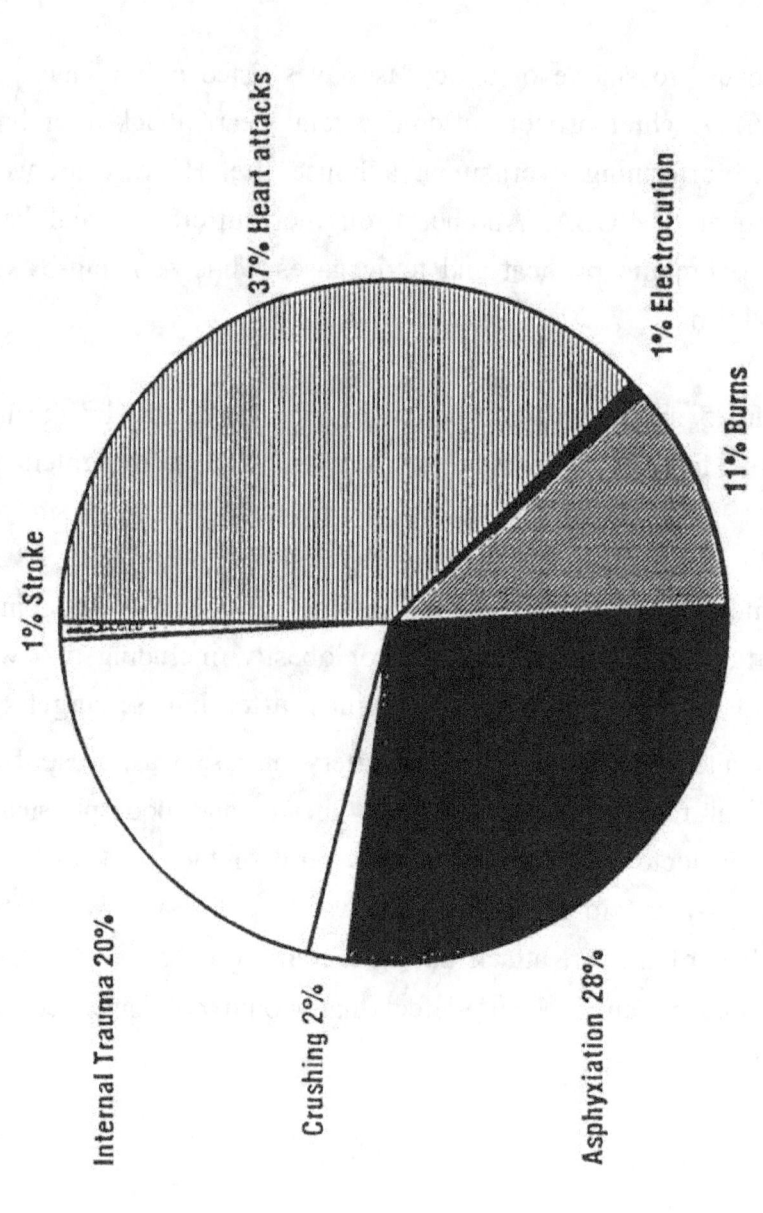

Figure 5
Firefighter Deaths by Nature of Fatal Injury 1994

37% Heart attacks

1% Electrocution

11% Burns

1% Stroke

Internal Trauma 20%

Crushing 2%

Asphyxiation 28%

Asphyxiation was the second leading medical reason for firefighter deaths. A total of 29 firefighter deaths (29%) resulted from carbon monoxide poisoning and/or smoke inhalation: 15 from wildland fires[6] and 14 at structure fires. This includes 7 firefighters who were using SCBA, but depleted their air supplies.

Internal trauma was responsible for 21 deaths (20%). This total includes 16 who were involved in vehicle or aircraft accidents.

Burn injuries claimed the lives of 11 firefighters (11%). Five died of burns after being caught in flashovers or backdrafts, 3 died of severe burns from an aircraft accident, 2 died from burns after being overrun by wildfires, and one died of burns after being caught in an explosion while not wearing any protective clothing.

Two firefighter deaths (2%) were attributed to crushing injuries: one died when he was crushed by falling debris from a roof collapse and another was crushed by a falling tree. One firefighter died when he was electrocuted (1%). Another firefighter died of a stroke (1%).

Ages of Firefighters

Figure 6 shows the distribution of firefighter deaths by age and cause of death. Younger firefighters were more likely to have died after becoming caught or trapped during firefighting operations. Stress was shown to play an increasing role in firefighter deaths as ages increased. Figure 7 shows the distribution of deaths by age and nature, with asphyxiation being the primary medical nature of death among younger firefighters, and heart attacks becoming much more prevalent among older firefighters. Twenty of the 27 firefighters who were over 50 years old and all 6 of the firefighters over 60 years old died from heart attacks.

---

6 This includes the 14 Storm King Mountain fatalities. The autopsy results established that their deaths were caused primarily by asphyxiation, secondarily by burns.

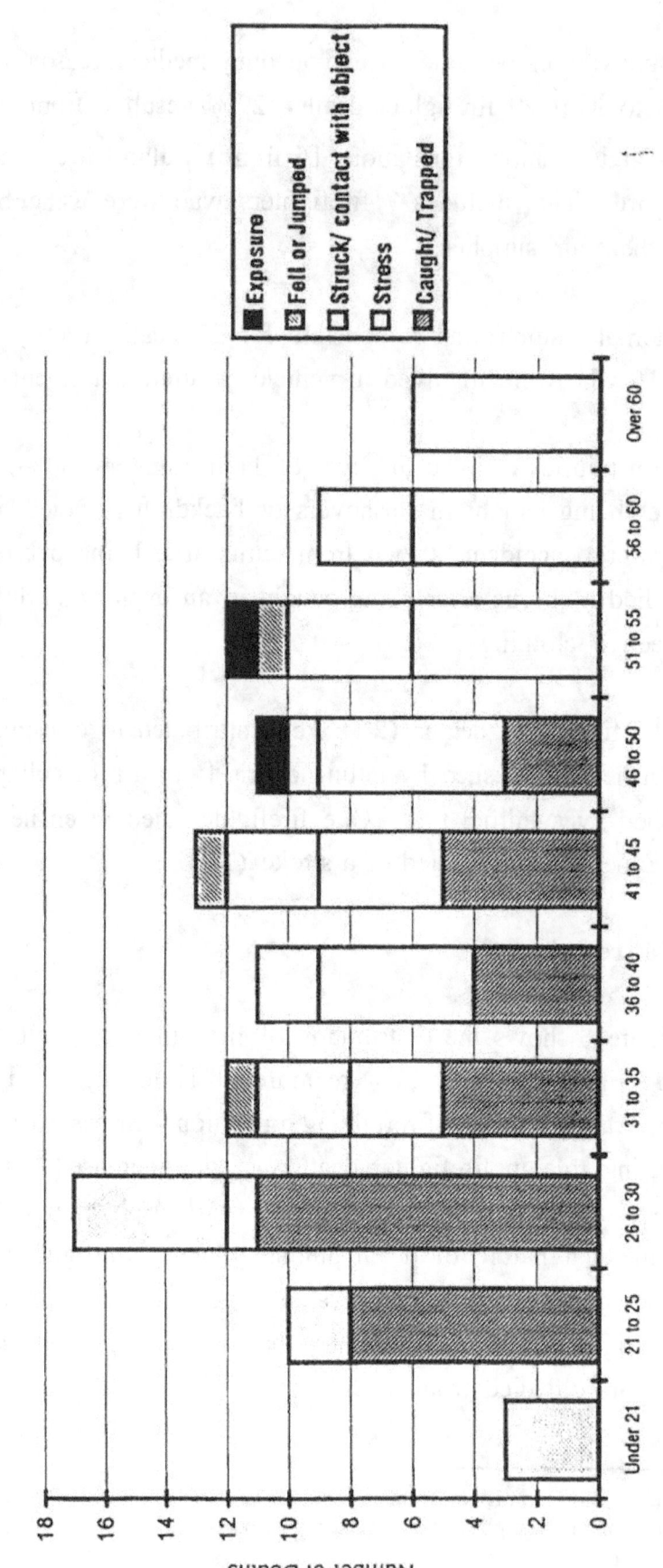

Figure 6

Firefighter Deaths by Age and Cause -1994

19

Figure 7

Firefighter Deaths by Age and Nature - 1994

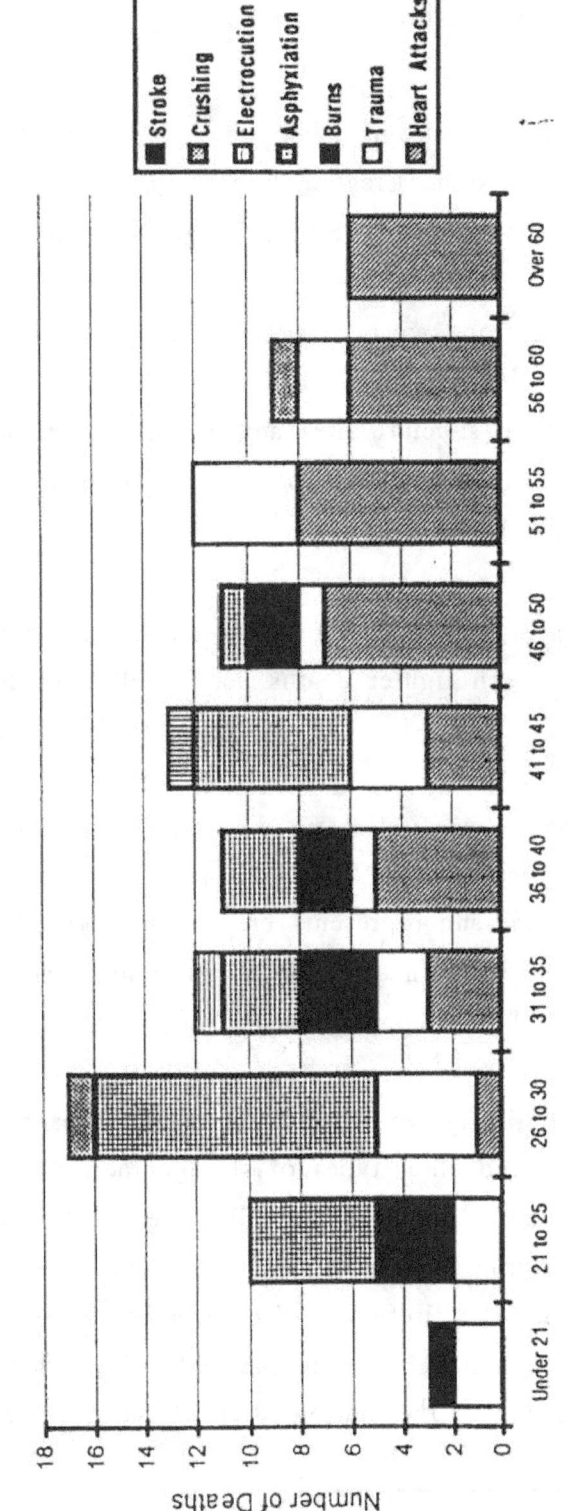

Fireground Deaths

Fireground deaths increased by 77 percent from 1993 to 1994.[7]
Figure 8 shows the fireground deaths by fixed property, including the 14
deaths that occurred on Storm King Mountain.

As in most years, residential occupancies accounted for the highest
number of fireground fatalities with 25 deaths. Residential fires account for
70-80% of all structure fires and a similar percentage of the civilian fire
deaths each year, but only 41% of the firefighter deaths in 1994 occurred in
residential structures[8].

Figure 13 shows firefighter death rates for property type. The data
indicates 5.5 firefighter deaths per 100,000 residential structure fires and 6.7
deaths per 100,000 fires in non-residential structure fires.

Twenty-two firefighters died while engaged in brush or wildland
firefighting in 1994, up from 8 in 1993. This is an increase of 175 percent
over last year and represents an estimated rate of 3.8 firefighter deaths per
100,000 wildland fires. Wildland firefighting is addressed as a special topic
at the end of this report.

Five firefighters died in storage occupancies, which include
warehouses and other types of storage facilities. Three firefighters died in
public assembly occupancies – all three in church fires. Two firefighters died
in manufacturing properties. One firefighter died in a fire in a commercial
use building, one firefighter died of a heart attack while standing by at an
open fire in a field, and one firefighter died when he suffered a heart attack
after returning to the station after an automobile fire.

---

7 Even if the 14 fireground fatalities from the Storm King Mountain incident were removed, the increase
would still be 35 percent.

8 Complete NFIRS data for 1994 fire incidence was not available at the time of this report, but typically
residential fires account for between 70 and 80 percent of all fatal fires each year.

21

Figure 8

Fireground Deaths by Fixed Property Use - 1994

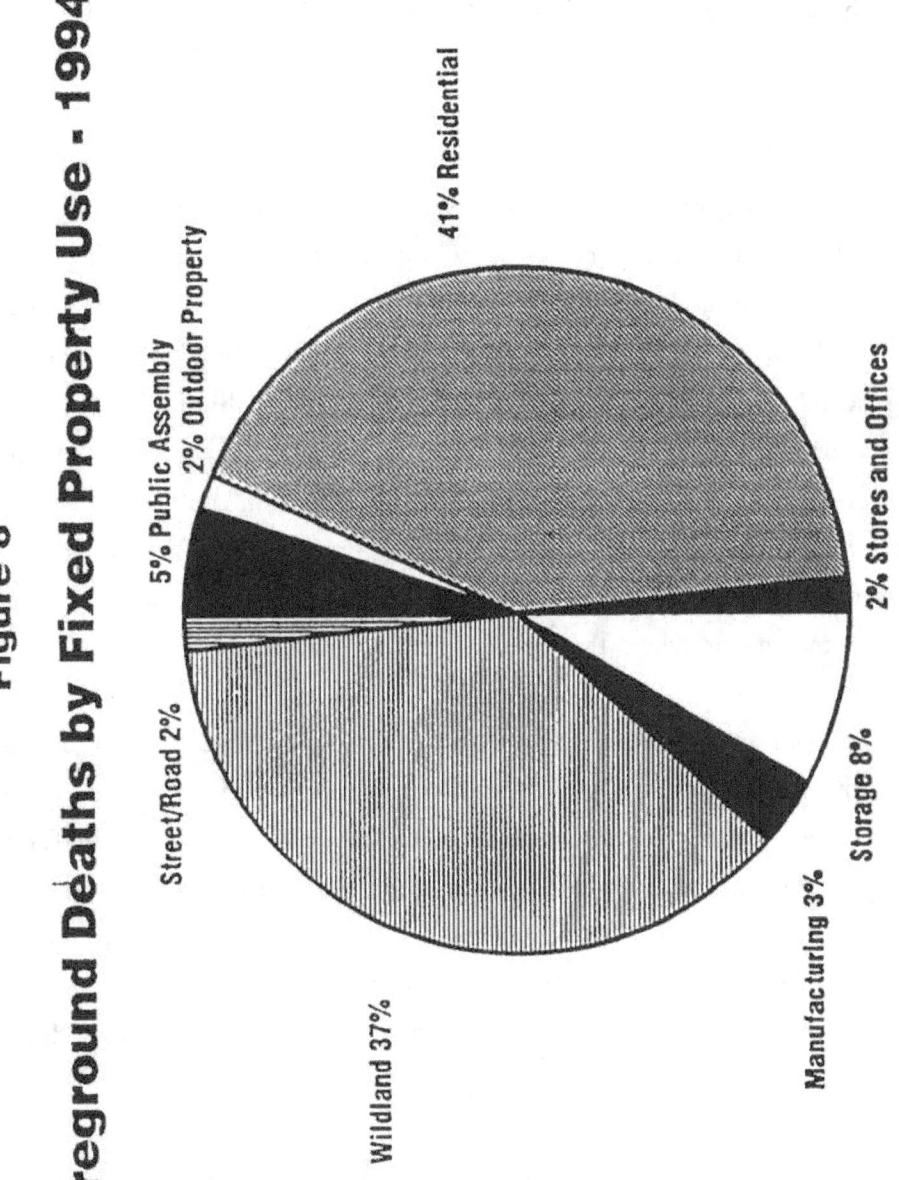

41% Residential

5% Public Assembly

2% Outdoor Property

2% Stores and Offices

Street/Road 2%

Storage 8%

Manufacturing 3%

Wildland 37%

Figure 9 illustrates the activity the deceased firefighters were engaged in at the time they sustained the injury or illness that caused their death. The activity of cutting fire lines to contain grass, brush, and forest fires accounted for 19 deaths, which is 32 percent of all fireground fatalities and reflects the high number of wildland fatalities in 1994.

Search and rescue operations in burning structures were being conducted when 14 deaths occurred. This activity was associated with 23 percent of the 60 fireground deaths and was the second highest category overall, and the highest category for structure fires. Analysis of these deaths reveals that 9 died of asphyxiation, 4 died of burns, and one died of a heart attack. At least 12 of these deaths may have involved firefighters who were conducting search operations on or above the fire floor while they or other companies were also ventilating the structure, resulting in rapidly changing fire conditions. This indicates the need for coordination between ventilation, search, and attack on all fires, and emphasizes the dangers of conducting search and rescue operations without a protective hose line, whether the fire location is unknown, or when the fire has not yet been confined.

Support and other duties on the fireground accounted for 10 deaths or 17 percent of the 1994 fireground fatalities. (Support duties include utility control, laddering, setting up equipment, forcible entry, or who were engaged in dropping retardant on a wildfire when their plane crashed, and an industrial firefighter who died when he ran out of air at a fire in a building at a mine. Several firefighters suffered fatal heart attacks after arriving on the fireground, but before they could be assigned to specific activity and they are included in this category.

Performing water supply activities resulted in 5 heart attack deaths (8 percent of fireground deaths). Three of these individuals were operating fire pumps and two were engaged in pulling supply lines.

23

Figure 9

Fireground Deaths by Type of Activity - 1994

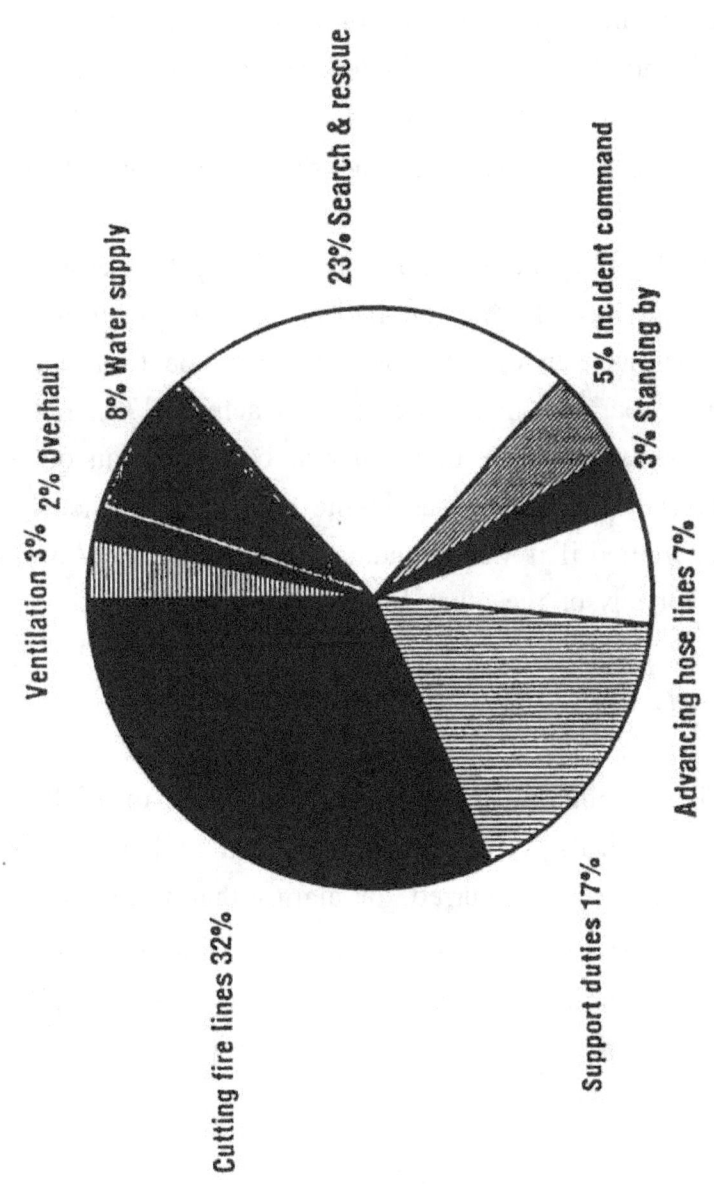

Ventilation 3%

2% Overhaul

8% Water supply

23% Search & rescue

5% Incident command

3% Standing by

Advancing hose lines 7%

Support duties 17%

Cutting fire lines 32%

The remaining twelve deaths (20 percent of the fireground deaths) involved firefighters performing various other firefighting activities. This includes 4 firefighters who died while advancing hose lines (7%), 2 officers who suffered heart attacks and one officer who was burned while they were commanding incidents (5%), 2 firefighters performing ventilation (3% – one fell through the roof, while the other was electrocuted descending an aerial ladder). In addition, two firefighters (3%) died of heart attacks while standing by at controlled burns and one chief officer (2%) died during overhaul when he was exposed to toxic gases and suffered a heart attack.

Of the fireground deaths where firefighters were caught or trapped in burning buildings, 4 were reported as having been found wearing PASS devices that were in the off position. Only one firefighter was reported to have been wearing a PASS device that activated when he became trapped, but rescuers were unable to reach him before he ran out of air. Another firefighter's PASS device was in the armed mode when he was found, but it was not reported if it was sounding. The status of PASS devices was not reported for any of the other fireground fatalities.

Time of Alarm

The distribution of 1994 fireground deaths according to the time of day when the incidents were reported is shown in Figure 10. The highest number of fireground deaths occurred for alarms that were received between 3 PM and 5 PM.[9] The times of alarms where firefighters died fairly evenly distributed, with higher rates at 1 PM to 3 PM, 3 PM to 5 PM, and 7 PM to 9 PM.

---

[9] The high total is driven by the 14 firefighters killed on Storm King Mountain; the blow up occurred just before 5 PM, but the fire had been burning for several days.

25

Figure 10

Firefighter Deaths by Time of Alarm - 1994

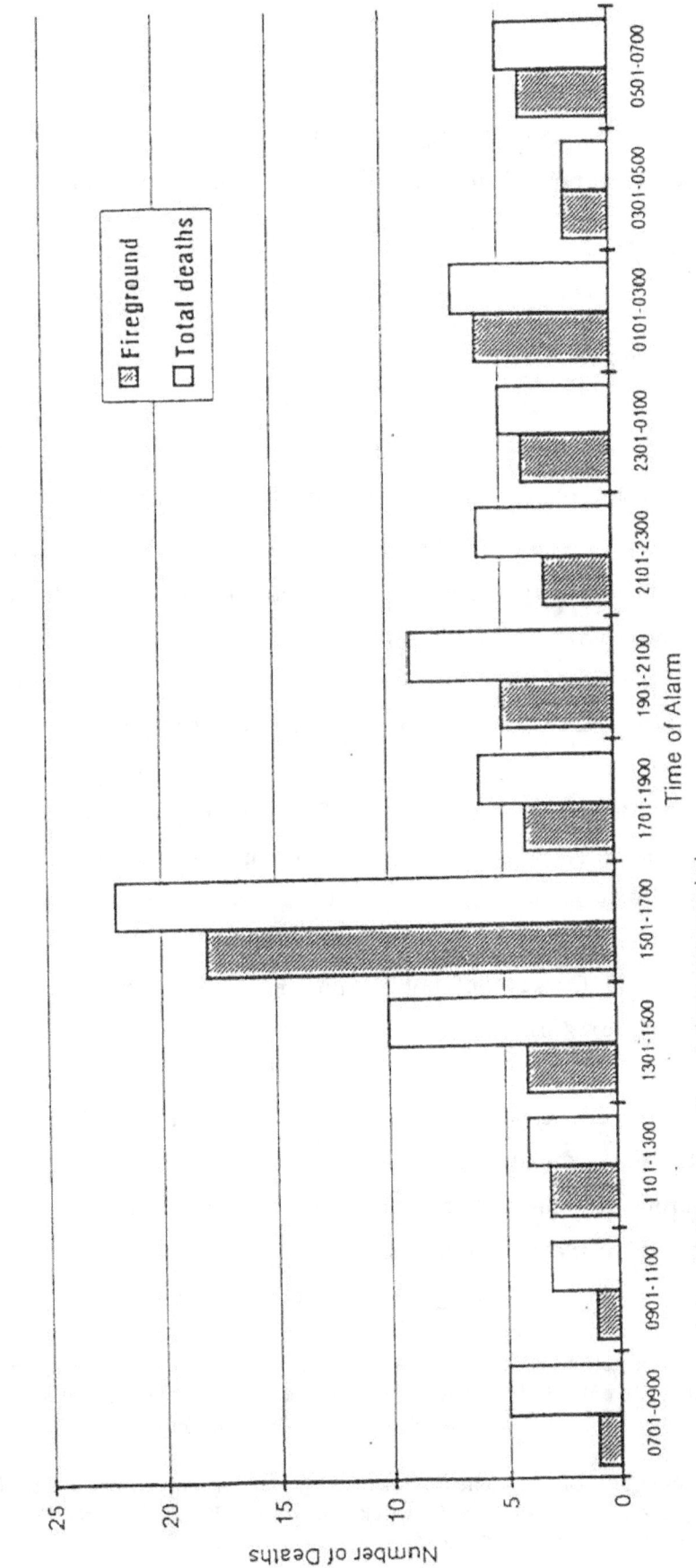

Based on 88 firefighter fatalities where time of alarm was reported.

Month of the Year

Figure 11 illustrates firefighter fatalities by month of the year. Firefighter fatalities peaked in July, the month of the Storm King Mountain fire. Other high months were recorded in January and August May had the lowest number of deaths.

State and Region

The distribution of firefighter deaths by state is shown in Table 1.[10] Thirty-five states are represented on the list, led by New York with 14 deaths.[11] Figure- 12 show the firefighter fatalities divided by region of the country and whether they were career structural, volunteer structural, or career or seasonal wildland firefighters.

Analysis of Urban/Rural/Suburban Patterns in Firefighter Fatalities

The U.S. Bureau of the Census defines urban as a place having at least 2,500 population or lying within a designated urban area. Rural is defined as any community that is not urban. Suburban is not a census term but may be taken to refer to any place, urban or rural, that lies within a metropolitan area defined by the Census but not within one of the designated central cities of that metropolitan area.

Fire department areas do not always conform to the boundaries of Census places. For example, fire departments organized by counties or special fire protection districts may have both urban and rural sections, and federal, state, and private firefighters. In such cases, it may not be possible to characterize the entire coverage area of the fire department as rural or

---

10 This list attributes the deaths according to the state where the fire department or unit is based as opposed to the state where the death occurred. They are listed by those states for statistical purposes, and for the National Fallen Firefighters' Memorial at the National Fire Academy.

11 The South Canyon Fire which occurred at Storm King Mountain **In Colorado claimed 14 lives, however the** table lists only one of these individuals from Colorado; 2 were from Idaho, 1 **was** from Montana, 1 **was** from South Carolina, and 9 were from Oregon.

Figure 11

Firefighter Deaths by Month of Year - 1994

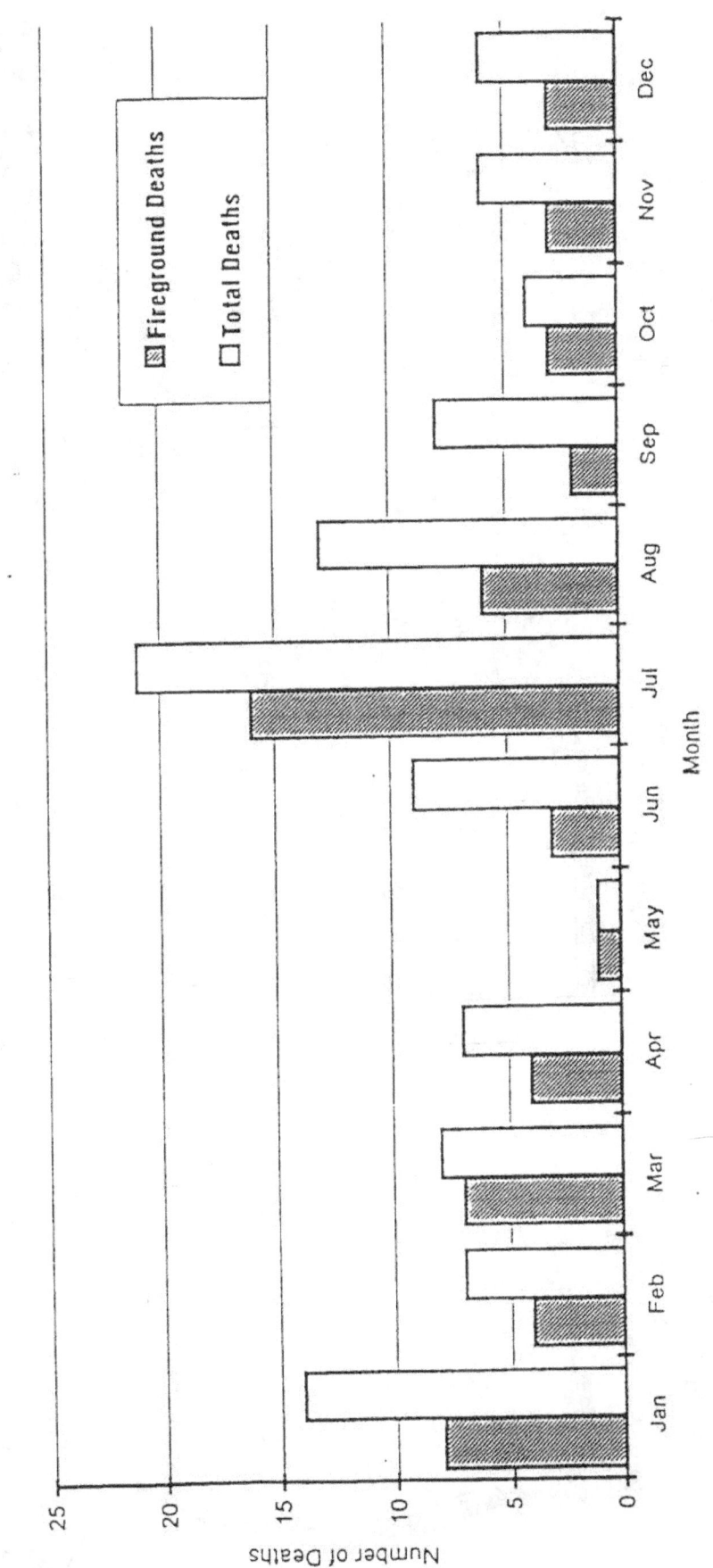

Figure 12

Analysis Report on Firefighter Fatalities in the
United States in 1994

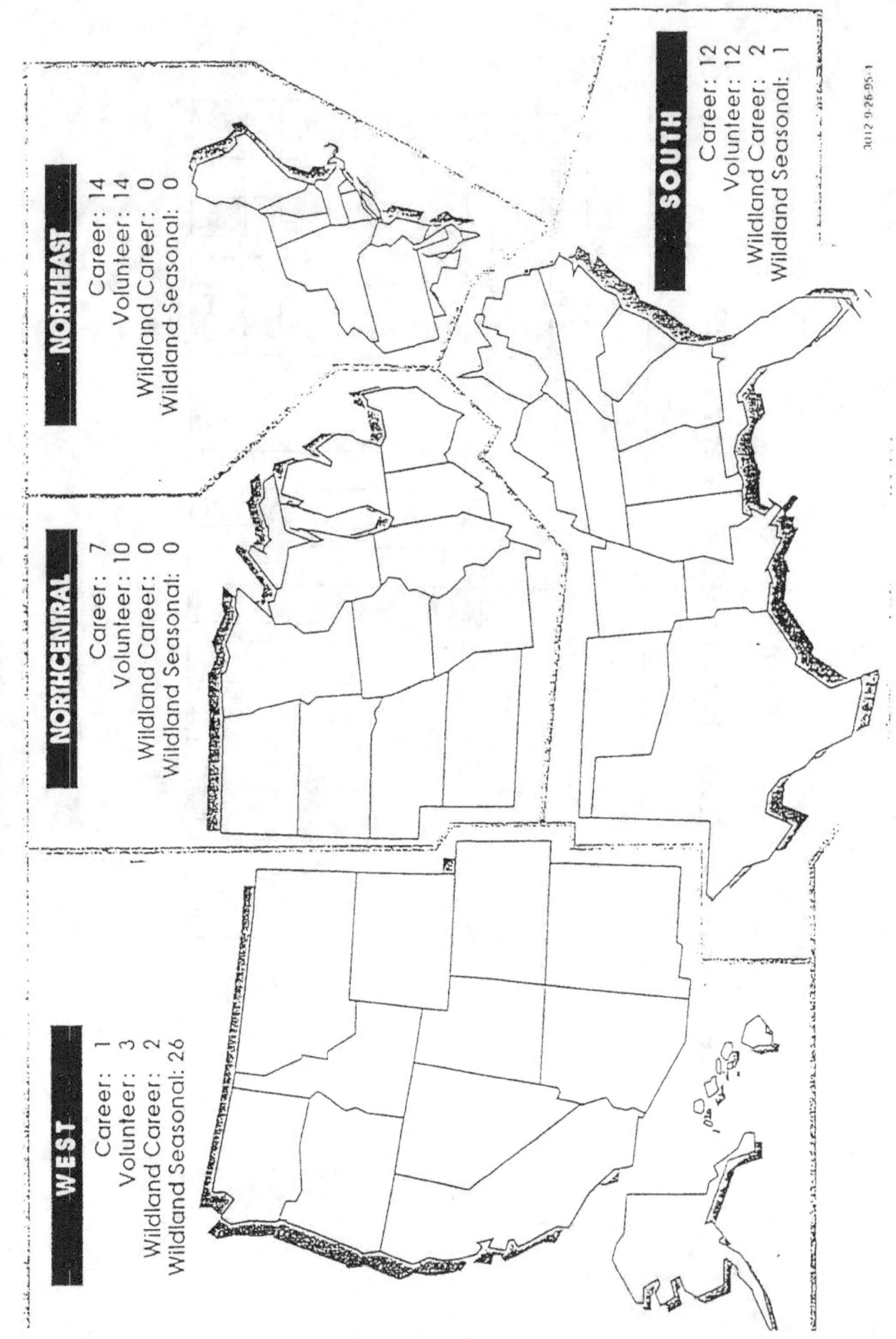

NORTHEAST
Career: 14
Volunteer: 14
Wildland Career: 0
Wildland Seasonal: 0

NORTHCENTRAL
Career: 7
Volunteer: 10
Wildland Career: 0
Wildland Seasonal: 0

WEST
Career: 1
Volunteer: 3
Wildland Career: 2
Wildland Seasonal: 26

SOUTH
Career: 12
Volunteer: 12
Wildland Career: 2
Wildland Seasonal: 1

Table 1
1994 On-Duty
Firefighter Fatalities

State	Number of Deaths	State	Number of Deaths
Alabama	3	New Jersey	7
Arizona	2	New Mexico	4
Arkansas	1	New York	14
California	5	North Carolina	1
Colorado	1	Ohio	4
Connecticut	1	Oklahoma	4
Florida	2	Oregon	10
Georgia	1	Pennsylvania	3
Idaho	4	Rhode Island	1
Indiana	3	South Carolina	2
Kentucky	2	Tennessee	4
Maryland	1	Texas	3
Massachusetts	1	Vermont	1
Michigan	2	Virginia	2
Missouri	4	Washington	2
Montana	3	West Virginia	1
Nebraska	1	Wisconsin	3
Nevada	1		

Total: 104

urban, and one must assign a firefighter death as urban or rural based on the particular community in which the fatality occurred.

The following patterns were found for 1994 firefighter fatalities.

	Urban	Suburban	Rural	Federal or State Parks	Total
Firefighter Deaths	28	19	26	31	104

Past analysis of urban, suburban and rural firefighter fatalities has shown that little correlation exists between the demographic nature of a department's area and firefighter fatalities. Experience has shown that training, use of proper equipment, and incident management have more of an impact on firefighter safety than geographic location.

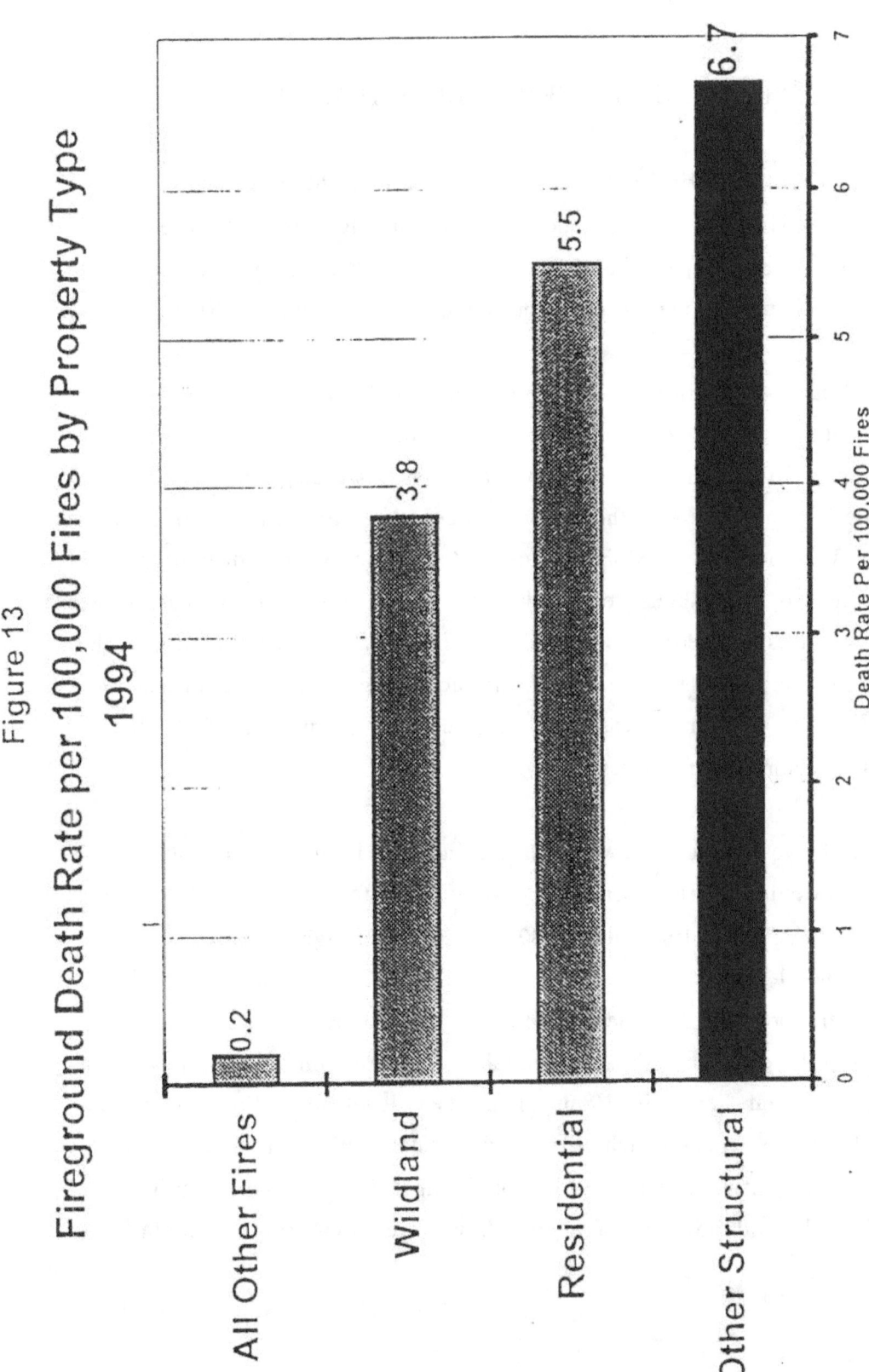

Figure 13

Fireground Death Rate per 100,000 Fires by Property Type
1994

Based on the number of fires as reported in the NFPA 1994 National Fire Experience Survey. NFIRS data for 1994 fire incidence was not available at the time of this report.

WILDLAND FIREFIGHTER FATALITIES

Thirty-eight (38) firefighters died in 1994 in wildland-related activities. Thirty-six (36) of these deaths were directly related to fire incidents; while two were wildland firefighters who died in job-related accidents. Twenty-two of the deaths occurred on the fireground (Figure 14).

The deaths of 14 wildland firefighters on July 6, 1994 in a single incident on Storm King Mountain in Colorado has drawn much attention to the subject of wildland firefighter safety, and has resulted in critical reviews of the particular incident and the circumstances that resulted in a tragedy of this magnitude. There were 8 fireground fatalities at other wildland fire incidents, which is fairly consistent with most years in the in the past decade. The toll of 14 lives lost enroute to wildland fires is also unusually high. The number of wildland deaths appears to vary according to the overall amount of fire activity in the wildland areas each year and the occurrence of single incidents that claim several lives.

While the loss of 14 lives in a single incident is an exceptional event, it is not unprecedented. The Storm King Mountain incident is the largest death toll in a single incident since 1937, when 15 firefighters died in the Shoshone National Forest in Wyoming. More recently, 7 died in a fire in the Tonto National Forest in Arizona in 1990, the most recent year with a high number of wildland deaths. Previous notable incidents include 12 deaths in the Angeles National Forest in 1966, 11 in the Cleveland National Forest in 1956, and 13 at the Mann Gulch Fire in the Helena National Forest in 1949. Each of these incidents involved a group of firefighters who were caught by an unanticipated rapid advance of a fire they were attempting to contain.

33

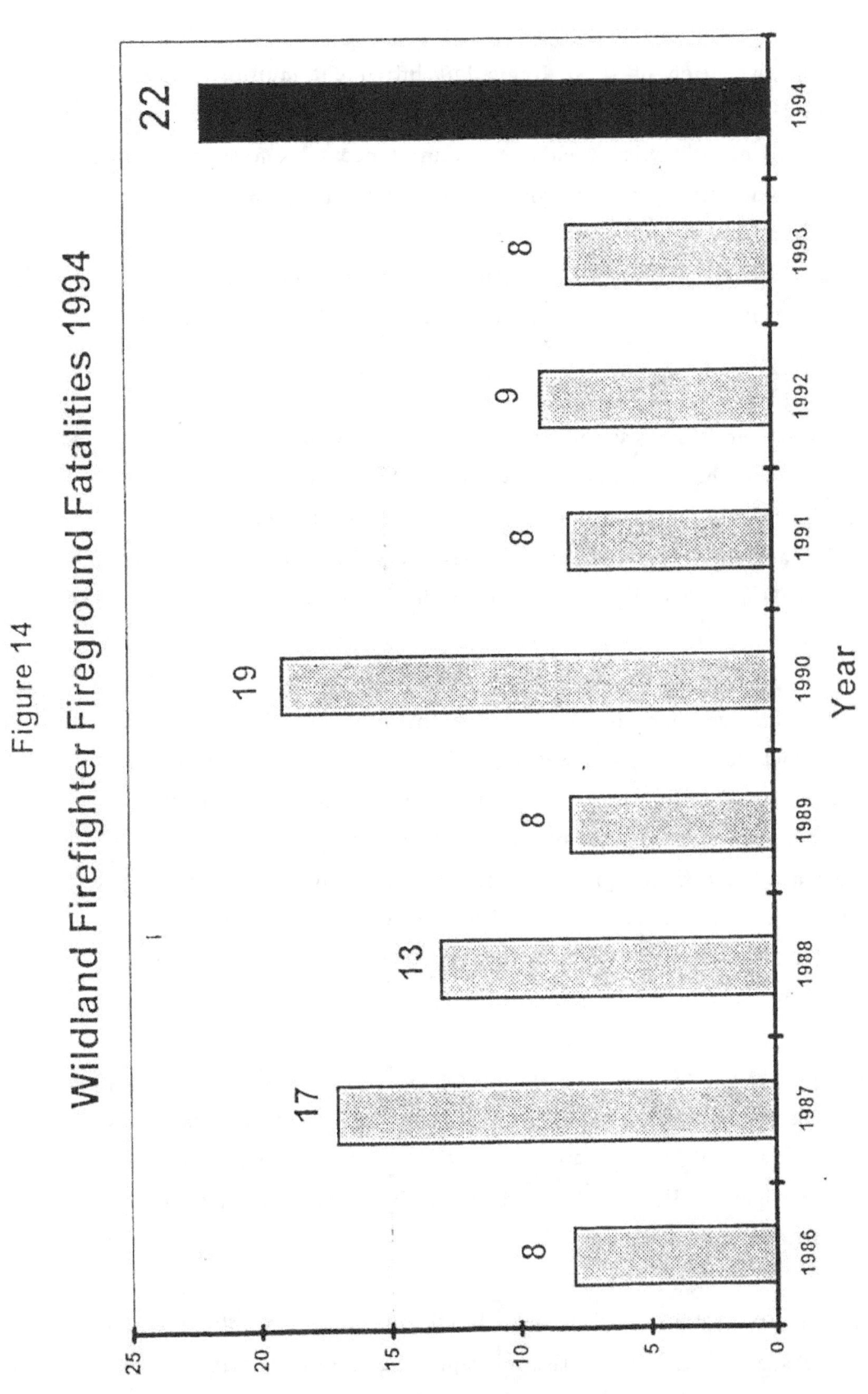

Figure 14

Wildland Firefighter Fireground Fatalities 1994

With the exception of these tragic but infrequent multiple death incidents, most wildland firefighter deaths appear to be individuals who are killed as a result of particular events or circumstances. In most years, there are a few individual firefighters caught or trapped by flare-ups, a few struck by falling trees, a few killed in vehicle rollovers, one or two struck by lightning, and a small number fatally injured in plane crashes. While the actual numbers vary from year to year, there are a few isolated incidents of each type most years.

The most significant variable with respect to the number of deaths each year appears to be the number of wildland fires that occur, which has a distinct relationship to climatic conditions. There may be additional factors, such as fuel management policies and policies on where and when to commit resources to fight fires which also impact on the death rate, however, there is no data available on these factors.

In 1994, the weather conditions were unusually dry in the western United States and the fire season was described as the worst on record in terms of the number of fires and, particularly, the number of large fires that occurred. These fires caused tens of thousands of firefighters to be deployed to an estimated 74,000 fires over a five-month period, including more than 100 that were described as major fires. The number of fires and the commitment of resources to attempt to control them were estimated as four to five times greater than most years.

Previous years with notably high numbers of wildland firefighting deaths were also years with elevated fire severity and frequency experience, particularly 1990 and 1987; these are the only years since at least 1978 in which more firefighters died fighting wildland fires than died fighting residential structure fires.

The hot, dry conditions that make wildlands more susceptible to the outbreak of fires also increase the risk of rapid fire spread and the danger of a sudden flare-up. All of the large life loss incidents appear to have occurred during high fire risk years. The exposure factor is a very significant variable when any attempt is made to determine trends or rates of fatal incidents in wildland firefighting. The increased number and size of fires in a busy season brings out many more firefighters and places them on the line for longer periods. It is not unusual, in a busy year, for crews to spend a week or more on one fire and then to be immediately assigned to another fire, possibly hundreds of miles away, with little or no rest. The time they are exposed to the risks is greatly increased, along with the fatigue factor, and the fire risk conditions are likely to be extremely high.

In addition to increasing the duration of exposure for individual firefighter and crews, the high fire activity years cause the number of personnel and units in action to be radically increased. The great majority of wildland firefighters are seasonally employed in a variety of categories. High activity periods cause the better trained, more experienced crews to be supplemented by less experienced crews, and ultimately by resources that are quickly organized and given only the most basic training before being committed to fires. In many cases, military resources, municipal firefighters, prisoners, and other sources of labor are deployed to supplement the wildland fire forces.

The process of expanding and deploying wildland firefighting resources has become much better organized in the recent past, however, it is still very difficult to reasonably estimate the number of personnel who are engaged in fighting wildland fires or the duration of their exposure to front-line firefighting risks. There are hundreds of different agencies and organizations and thousands of individuals and units of different types involved in this system at the federal, state, and local levels. There are records available to track the commitment of federal and most state resources, however it is very difficult to determine how much time each unit was actually involved in fire control activities or the level of risk presented by each situation. There is no system to closely monitor the commitment and use of local resources.

While the Storm King Mountain incident is a singularly tragic event, it should be noted that only seven wildland firefighters died while operating on all of the other wildland fires that occurred in 1994. Seven fatalities would be fairly consistent with most other years and may be low when the combined factors of exposure and fire severity are considered

This analysis includes several different categories of firefighters and different types of fires. It is very difficult to make valid comparisons between the number or frequency of firefighter deaths in wildland operations and structural firefighting operations. Structural firefighters routinely engage in very close contact with a fire they are trying to control – often within the same room – and are well protected from very short duration exposures to the fire by their protective clothing. Interior operations are usually conducted with self-contained breathing apparatus that protects the user while limiting the duration of an entry to relatively short periods. Very few structural fires have a duration of more than one or two hours. In contrast, wildland firefighters wear very little protection from the fire or the products of combustion and must use distance from the fire and avoidance of the smoke as their primary personal protection. A wildland firefighting operation may extend over several days or even weeks with long work periods between rests breaks. When faced with a rapidly advancing fire front, the wildland firefighter generally has only two choices – try to outrun it or find an area to deploy a personal shelter and hope that it provides adequate protection for the severity and duration of the direct fire exposure.

The magnitude of the fire in relation to available fire suppression capabilities is often very different in wildland fires than in structural fires. An urban fire department can often apply a sufficient water flow to control the fire, while a wildland firefighter can only attempt to create a fuel break ahead of the fire to stop its advance. Wildland situations often involve much larger fires and negligible fire suppression capacity. If conditions cause the fire to suddenly advance toward the firefighters, they usually do not have any capability to counterattack with an extinguishing agent. In a crisis situation, fleeing structural firefighters can often escape to a safe area within less than a minute, while the wildland firefighters may be several minutes or even miles from a safe refuge.

As noted in previous discussions of wildland fatalities, the causes of death for wildland firefighters tend to involve a much higher percentage of burns, asphyxiation, and trauma and a much lower percentage of stress related deaths than career and volunteer municipal departments. Only five of the 38 deaths related to wildland firefighting were caused by heart attacks and four of the five heart attack victims were members of local volunteer fire departments, as opposed to full-time or seasonally employed wildland firefighters. This reflects a significantly higher level of physical fitness and conditioning which the major wildland firefighting agencies have required of their personnel for several years. Because they employ large numbers of seasonal firefighters, the workforces of wildland organizations tend to be younger than most career or volunteer fire departments.

Wildland firefighting may involve both career and seasonal, temporary, or contract personnel working for various state and federal agencies that provide fire protection to different jurisdictional areas. In addition, career and volunteer firefighters who are associated with urban or rural fire departments may be called to respond to wildland fires or fires in urban/ wildland interface areas. The 1994 wildland firefighting deaths included 17 seasonal employees of federal wildland agencies, 14 at Storm King Mountain and 2 in a helicopter crash in New Mexico. Another seasonal employee died when a water tender rolled over enroute to a fire, and 2 career U.S. Forest Service employees died in other fire related incidents. Seven contract employees involved in wildland operations also died in four separate aircraft accidents.

There were no career firefighter deaths involved with wildland firefighting operations. Seven volunteer firefighters died in separate incidents while performing duties related to wildland firefighting.

---

12 The deaths at Storm King Mountain included 3 smokejumpers, 2 helitack crew members, 2 hot shot crew leaders, and 7 hot shot crew members.

Cause of Death

Seventeen wildland firefighters died in situations where they were caught and burned by quickly moving fires that cut off their escape routes. The direct cause of deaths for all the 14 firefighters who died at Storm King Mountain was determined to be asphyxiation; however, they were also severely burned by a rapidly advancing fire that raced up the side of a steep slope. In addition to the

14 seasonally employed wildland firefighters at Storm King Mountain, a female USFS career firefighter, a swamper (bulldozer operator) working as a contractor for a state forestry agency, and one member of a volunteer fire department that had responded to a large grass fire all died in similar situations.

Aircraft incidents associated with wildland firefighting caused a total of 9 deaths during 1994. The pilot and co-pilot of one aircraft died when it crashed while dropping retardant on a forest fire, and the 3 members of another air tanker died when it crashed enroute to a fire. Both air crews were employed by private companies that contracted with the United States Forest Service. Another contract pilot and 2 of the 4 USFS helitack crew members on board perished when their helicopter crashed enroute to a fire. A civilian employee of a U.S. Army helicopter unit that had been mobilized to support wildland firefighting operations died when he was struck by a rotor blade in a landing zone accident.

Aircraft incidents have been a major cause for concern in previous years, particularly with respect to the conditions of the aircraft and crew fatigue. In a busy fire season aircraft are heavily utilized and take on the characteristics of military flying operations, with long hours and fast turn-around between missions as the pilots fly dozens of sorties. Significant efforts have been directed toward upgrading the condition and safety of the aircraft used for these missions and providing adequate rest for the pilots. Although the loss of 3 aircraft and 8 lives does not appear to be an improvement over previous years, the loss rate appears to have improved significantly in proportion to the very high number of sorties that were flown in 1994.

Vehicle accidents enroute to wildland fires took the lives of 4 firefighters; 2 when their vehicles rolled over, one when a brush truck drove over an embankment, and one when a supply truck was involved in a collision while enroute to a fire.

Two career forest service firefighting personnel lost their lives in accidents that occurred while they were performing duties not directly related to their firefighting responsibilities. One died when a tree fell on him while a crew was clearing a path for a road. A second death resulted from an explosion that occurred while a fire management officer was being trained to fire a recoilless rifle used in avalanche control.

Heart attacks claimed the lives of 5 of the 38 firefighters listed as having died in wildland related incidents in 1994. One volunteer firefighter died while fighting a brush fire and one died while returning to his vehicle after responding to an incident that had been reported as a grass fire that turned out to be a legal camp fire. Another volunteer died while operating a road grader to plow a firebreak around a grass fire. A seasonally employed wildland firefighter suffered a fatal heart attack while driving a water tender enroute to a fire. A career district ranger died of a heart attack while monitoring a prescribed burn.

There was only one heart attack fatality among the thousands of firefighters who were mobilized to fight the thousands of wildfires that broke out in the Western United States in the summer of 1994. This reflects well on the mandatory requirements for medical examinations and physical fitness that have been enforced by the major wildland fire suppression agencies. It also underlines the factor of age in heart attacks, since the seasonally employed wildland firefighters tend to be much younger than the general population of structural volunteer and career firefighters.

RISK MANAGEMENT

A major concern has been identified with the concept of operational risk management in relation to firefighting and other emergency services in the past two years. This concern is most directly expressed in NFPA 1500, Standard for Fire Department Occupational Safety and Health Program (1992 Edition), which requires a "risk management approach" to be employed to determine the amount of risk that is acceptable in different emergency situations. In effect, firefighters and fire officers must make a common sense approach to all operations, evaluating each situation for the level of risk involved, and realizing in what situations personnel may need to take risk and in which situations that they should not. The objective of this requirement is to avoid exposing firefighters to excessive risk of injury or death, particularly where it serves no justifiable purpose.

There is currently no accepted scale or standard to measure risk, and most evaluation of what is acceptable usually takes place after a death, injury, or close call has occurred. It is evident in many firefighter death investigations that there is usually very little recognition of the risks before something went wrong. Many situations appear to have been low risk or "routine" until something went wrong and it was too late to prevent a fatality.

There is an expression, "If it can be predicted, it can be prevented," and a post incident analysis version, "If it could *have been* predicted, it could *have been* prevented." The issue that needs to be addressed, in order to refine risk management, is how to recognize situations that should cause emergency responders to predict danger and take steps to reduce or avoid those risks. This approach should be adopted as a standard process in the analysis of a firefighter fatality. The fire service should adopt it as a standard part of the initial and ongoing size-up of all its emergency incidents.

The 1994 firefighter fatalities present numerous familiar scenarios that have occurred in previous years. It is impossible to determine the state of mind of each individual who died or who was in a position to influence the event; however, it is feasible in many cases to determine if there were risk indicators that could have, or should have, been recognized – or if there were no indications of significant risk.

An attempt was made to look for indications of risk recognition or consideration of alternative approaches in the analysis of 1994 fatalities records, however, no documentation of this factor could be found in the information that was obtained on any of the 104 deaths. This information has not been collected or reported as part of the existing system for tracking firefighter deaths.

It appears that most of the deaths occurred in situations that were not perceived or recognized as high risk. There was information revealed in the investigation of the Storm King Mountain incident that some of the individuals had discussed the fact that they were operating in violation of safety procedures, which suggests that they failed to recognize the risks that resulted from these actions. Further analysis of this subject will require a more structured approach to obtain the information. However, there has been no organized effort to obtain this information.

The most evident contradiction in this analysis is the number of heart attack deaths that occurred in individuals who were known to have previous cardiac problems, significant obesity, and/or poor physical conditioning. In spite of these factors, these individuals were involved in emergency response functions that could be predicted to involve high stress and physical exertion, where their conditions posed a predictable and identifiable risk to themselves and others.

Cases where the risk identification would depend upon the presence and recognition of some other indicators are more difficult to document. This challenge is not adequately addressed through current methods of recording and reporting firefighter fatalities. This subject should be incorporated into the analysis in the future.

CONCLUSIONS

The analysis of firefighter deaths for 1994 provides a mixture of good news and bad news. The increase of 35% in the number of on-duty fatalities over the two previous years also caused the total to exceed the benchmark of 100 for the first time since 1991. The number of fireground deaths increased over the previous years and the number of wildland firefighting deaths reached the highest level in the 17 years this analysis has been conducted for the U.S. Fire Administration. The loss of 14 lives in a single incident was the most tragic event in the worst year that has been recorded for wildland fires.

While the overall statistics are distressing, there are several encouraging signs within the analysis. Although they still account for a substantial number of deaths, the number of heart attacks appears to be decreasing, particularly in the wildland firefighting establishment. In 1994 there was only one death of a volunteer firefighter responding to an alarm in a private vehicle and deaths caused by falls from apparatus have been almost eliminated; both of these are problems that have been identified as needing attention in previous studies. Even with 104 deaths, the total for 1994 is still lower than any year prior to 1991.

Several areas can be identified from the study that require attention to prevent future line of duty deaths. The need for increased research and development relating to safety in the wildland environment is evident and has been widely recognized. Several advances have been made to provide and improve accountability systems for firefighters conducting interior operations at structure fires, however, there continues to be several lives lost each year when firefighters become lost or disoriented and run out of air. A problem has been identified with firefighters not activating their PASS devices when operating inside structures – four fatalities were specifically noted as having been found with their PASS devices in the off position, and only one was reported to have been found with a PASS operating. The functionality of current PASS units and advanced designs needs to be evaluated, as well as the training that is provided on their use.

The full analysis of firefighter fatalities also suggests that more attention needs to be directed toward operational risk management. Many of the individual cases suggest circumstances where the nature and degree of the risks involved do not appear to have been recognized or evaluated. There is a weakness in the current ability to document risk recognition and analysis factors which should be addressed in time to improve the analytical capabilities in this regard.

APPENDIX A: SUMMARY OF FATAL 1994 INCIDENTS

Incident 1

On January 1, Firefighter Ronnie Fuller of the Clinton (SC) Fire Department died after suffering a heart attack while operating a pumper at a house fire. His fellow firefighters and police officers on the scene immediately rushed to his aid when they saw him collapse and initiated CPR. He was transported to the hospital where he died.

Incident 2

On January 3, Firefighter Marcus Carr and Firefighter/Paramedic David Mosher, both of the Chillicothe (MO) Fire Department were killed when their ambulance was struck head-on by a tractor-trailer truck that veered into their lane. The two were enroute to the hospital with a patient, who also died in the crash. The ambulance was running with lights and siren at the time of the accident. The truck driver was critically injured.

Incident 3

On January 4, Firefighter Thomas Dunn of the Rutherford (NJ) Fire Department was killed when he became trapped on the second floor by a rapidly advancing fire in a balloon frame house. Dunn's company had been conducting search operations and horizontal ventilation in the house with only light fire and smoke conditions. Fire had initially been located in the basement. Conditions rapidly deteriorated as the pressurized heat and smoke broke out of concealed spaces in the attic above the fire fighters and from a dropped tin ceiling on the floor below them. Two personnel with Dunn were able to escape through second floor windows as evacuation signals were sounded. Dunn became disoriented and entangled in a bed frame on the second floor and was unable to escape. A rescue team entered the second floor via a window and quickly reached him, but he died from acute smoke inhalation and burns. Dunn's facepiece was not on when he was found, his SCBA straps had failed, and his PASS device was in the off position. A New Jersey Division of Fire Safety report indicates that the balloon frame construction, ventilation simultaneously below and above hidden fire areas, and operations of fog pattern streams in the basement may have all contributed to the rapidly deteriorating conditions faced by the fire crews on the second floor.

Incident 4

On January 8, Firefighter Gerald Mullins of the Binghampton (NY) Fire Department had just finished a shift assigned to an EMS unit and left the firehouse when he collapsed from a heart attack across the street. Medical care was quickly administered but efforts to revive him were unsuccessful.

Incident 5

On January 10, Deputy Chief Harold Salisbury of the East Greenich Fire Department was incident commander at a structure fire in a metal product manufacturing plant. During the incident Chief Salisbury collapsed from a heart attack. Efforts to revive him were unsuccessful and he was pronounced dead after transport to the hospital.

Incident 6

On January 12, Firefighter Dennis Mullins, Jr., of the Mount Vernon (NY) Fire Department suffered a heart attack at a fire. He died of complications from that heart attack in 1995.

Incident 7

On January 21, Firefighter Glen Thorn of the Sea Girt (NJ) Fire Company #1 suffered a heart attack at the scene of a structure fire after arriving in his personal vehide and preparing to perform support functions at the exterior of the structure. His collapse was witnessed by other fire department personnel who administered emergency care. He was transported to the hospital and died nineteen days later.

Incident 8

On January 22, Captain George Ciliberto of the Ocean City (NJ) Fire Department died of a heart attack in his sleep while on duty at the fire station. Captain Ciliberto had run several emergency calls during the evening.

Incident 9

On January 23, Firefighter Maurice Wardwell, Jr. of the Proctor (VT) Fire Department collapsed and died of an apparent heart attack after arriving on the scene of a working fire.

Incident 10

On January 28, Captain Nick Charmello of the Kansas City (MO) Fire Department had just finished assiscing in the extrication of a patient from an automobile accident when he collapsed from an apparent heart attack. Firefighters and rescue crews on the scene rushed to his aid. He was pronounced dead after being transported to the hospital.

Incident 11

On January 28, Firefighter Walter Franks of the Pine Hill (NJ) Fire Department stayed at the fire hall to prepare coffee for firefighters out on a call. They returned to find him unconscious, having suffered a heart attack. Emergency care was initiated. Franks died several days later at the hospital.

Incident 1 2

On January 28, Firefighters Vencent Acey and John Redmond, both of the Philadelphia (PA) Fire Department, died when the became trapped and overcome by smoke by a rapidly moving fire in the basement of a church. Several firefighters re-entered the church against orders to rescue the firefighters, and were able to pull one of them from the basement. Eight other firefighters were injured, including several involved in the rescue efforts.

Incident 13

On February 1, Firefighter Marilyn Williams of the Keystone (OK) Volunteer Fire Department suffered a heart attack and died while pulling hose at a mobile home fire.

Incident 14

On February 5, Battalion Chief Robert English of the Detroit (MI) Fire Department suffered a fatal heart attack while directing crews during the overhaul. stages of an apartment fire. Children playing with matches are believed to have caused the fire.

Incident 15

On February 7, Firefighter Newt Morgan of the Poughkeepsie (AR) Volunteer Fire Department was driving a fire engine to a reported structure fire when he suffered a massive heart attack. The engine veered slowly off the road into a tree. Firefighter Morgan was found in cardiac arrest by firefighters responding behind him and is believed to have died before the accident occurred.

Incident 16

On February 11, Fire Engineer Timothy Hale of the Phoenix (AZ) Fire Department was killed when he and his partner were struck by a vehicle while unloading the stretcher from the rear of their rescue unit during an EMS incident. Hale received severe traumatic injuries and died at a trauma center the following day. The driver of the vehicle was intoxicated.

Incident 17

On February 20, Firefighter Ann F. Sheppard of the Venus (FL) Volunteer Fire Department was participating in search and rescue training when she suffered a fatal heart attack.

Incident 18

February 26, District Ranger Bedford Cash of the U.S. Forest Service, was conducting prescribed burning in the Tuskegee National Forest when he suffered a fatal heart attack.

Incident 19

On February 27, Firefighter Dennis Dearing, Jr., of the Auburn Hills (MI) Fire Department died when the floor collapsed under him while conducting operations at a house fire. Firefighter Dearing and two others had entered the house through the kitchen with a hose line to try to reach a fire located in the basement that had been burning for approximately 40 minutes. The officer in charge of the attack crew ordered them to evacuate the house due to the spongy feeling of the floor as they approached the basement stairs, but the floor collapsed beneath Dearing before he could escape. The fire was ruled incendiary in nature, with a high fire load of combustibles in the basement, contributing to the floor collapse.

Incident 20

On March 2, Firefighter Mark Mitchell of the Pawcatuck (CT) Fire Department died of carbon monoxide poisoning after being separated from his crew while conducting search operations on the second floor of a single family house. Mitchell and three other firefighters were attempting to rescue a victim reported on the second floor when a flashover occurred, separating the crew members. Three firefighters escaped with injuries. Mitchell was found unconscious on the second floor and died later. His blood carboxy-hemoglobin level was 24%. It is believed that there was a delay in over one hour before the fire department was called.

Incident 21

On March 5, Firefighter Charles Butchee of the Warren Community (OK) Fire Department died of a heart attack after being exposed to smoke and heat at an outside controlled burn. Firefighter Butchee had been standing by during the event.

Incident 22

On February 6, Lieutenant Walter Wade of the Miami Township Fire Department died of a heart attack after completing a search for victims at a house fire. An autopsy revealed that Lt. Wade had an enlarged heart and a genetic heart abnormality.

Incident 23

On March 22, Firefighter Gary King of the Grundy County (MO) Rural Fire Protection District suffered a heart attack while operating at a two alarm brush fire. King died after being transported to the hospital.

Incident 24

On March 22, Firefighter Dustin Mills of the Capron (OK) Fire Department died enroute to a wildfire that burned over 5,000 acres. Firefighter Mills died when the brush truck he was riding on overturned as it drove over a smoke obscured 15' embankment. Mills died of traumatic injuries at the scene; another firefighter received minor injuries.

Incident 25

On March 29, three firefighters trapped in the stairwell of a brownstone were burned when they were enveloped in fire while attempting to force their way through a heavy steel door to a second floor apartment. Captain John Drennan, Firefighter James Young, and Firefighter Christopher Seidenburg of the New York City Fire Department were conducting a search when the hot air and toxic gases that collected in the stairwell erupted into flames as other fire crews forced entry into the first floor apartment where the *fire* had originated. The fire exhibited characteristics of both a backdraft and a flashover. Firefighter Young, in the bottom position on the stairs, was burned and died at the scene. Firefighter Seidenberg and Captain Drennan were rescued by other firefighters. They were transported to a burn unit with third and fourth degree burns over 50 of their bodies. Seidenburg died the nest day. Drennan passed away several weeks later. The fire cause was determined to be a plastic bag left by the residents on top of the stove of the floor apartment.

Incident 26

On April 2, Firefighter Joseph Jay Boothe of the Pea Ridge (AL) Volunteer Fire Department was riding in the passenger seat of a 1971 Ford 1200 gallon tanker truck enroute to a brush fire that was threatening several homes and a church. Boothe was killed when the vehicle overturned heading into a sharp turn. The police report indicated that the vehicle was traveling at approximately 35 miles per hour in a 30 mph zone. The driver received minor injuries and reported that the brakes on the vehicle locked up heading into the curve. The vehicle had no scatbelts and Firefighter Boothe died of traumatic injuries received from being pinned under the vehicle.

Incident 2 7

On April 3, Firefighter Robert Waskiewicz of the Augusta-Bridge Creek (WI) Fire Department received fatal burn injuries when he was caught in a wind shift and overrun by a fast moving grass fire.

Incident 28

On April 7, Firefighter Ronald Carlson of the Blue Creek (NE) Rural Fire Protection District/ Lewellen Volunteer Fire Department was driving fire apparatus to a brush fire when the vehicle rolled over, killing him and injuring two other firefighters seriously.

Incident 29

On April 11, Lt. Michael Mathis and Private William Bridges of the Memphis (TN) Fire Department were killed when they became trapped and overcome by smoke during a fire on the ninth floor of a high rise building. Two civilians also died in the arson fire. Lt. Mathis became disoriented when he was caught in rapidly spreading fire conditions on the fire floor, burning him and causing his SCBA to malfunction. He found his way into a room on the ninth floor were he was later discovered by other fire crews with his SCBA air depleted. Private Bridges, aware that Lt. Mathis was unaccounted for after several unsuccessful attempts to contact him by radio, left a safe stairwell where he had been attempting to fix problem with his own SCBA. Investigators believe Bridges was trying to locate Lt. Mathis. Bridges became entangled in fallen cable TV wiring within a few feet of the stairwell, and died of smoke inhalation after depleting his SCBA supply. A Memphis Fire Department investigation found many violations of standard operating procedures by companies on the scene, including crews taking the elevator to the fire floor, problems with the incident command system and coordination of companies, operating a ladder pipe with crews still on the fire floor, and a failure of personnel, including Lt. Mathis and Private Bridges, to activate their PASS devices.

Incident 30

On April 15, Stanley Rhoads, a member of the Barrick Goldstrike Mine Emergency Response Team, was on his way to work when a fire broke out in a gold refinery building. After arriving, he was witnessed putting on his personal protective clothing and SCBA. Two hours later, members of a volunteer fire department that had responded to the fire found his body inside the fire building. He had apparently entered the structure independently and ran out of air inside the refinery. Commanders did not know he was on the fire scene until his body was removed. The initial fire attack was described as "hectic" to the Nevada State Fire Marshal that investigated the report. His death was attributed to smoke inhalation.

Incident 31

On April. 29, Fire-Police Officer Joseph Jarvis, Sr., of the Oceanside (NY) Fire Department was struck and killed by a vehicle while directing traffic at an emergency scene.

Incident 32

On May 30, Firefighter Alton Warren of the Baltimore City (MD) Fire Department was injured when he fell down the stairs at a fire, breaking his ankle. He died later of an embolism that developed from the injury.

Incident 33

On June 4, Senior Firefighter Anthony Covas of the Newport News (VA) Fire Department died after suffering a heart attack while on duty in his station.

Firefighter Covas, a twenty year veteran, had participated in a morning physical training exercise, and had gone to a separate room in the firehouse after eating lunch. Fellow firefighters found him in cardiac arrest a few hours later.

Incident 34

On June 5, Lieutenant George Lener of the New York City Fire Department collapsed from smoke inhalation and was found unconscious in the basement of a five story warehouse after a fire that required the response of more than 300 firefighters. Lt. Lener died seven weeks later without regaining consciousness. An suspected arsonist has been arrested a charged with starting the fire. Sixteen other firefighters were injured during the incident.

Incident 35

On June 10, Firefighter Victor Ruth III of the Neptune (PA) Fire Company #1 suffered a fatal heart attack while responding as part of an engine company to a med-evac standby.

Incident 36

On June 13, Private Marc Butcher of the Parkersburg (WV) Fire Department died in his sleep of a heart attack while on duty at the fire station. Medical efforts by fellow firefighters to revive him were unsuccessful.

Incident 37

On June 14, Firefighter Ron Holmgreen of the Lake Havasu (AZ) Fire Department suffered a fatal heart attack shortly after returning home from a fire department drill. Firefighter Holmgreen had exhibited signs of cardiac distress during the drill.

Incident 38

On June 18, Firefighter/EMS Coordinator David Barter of the West Terre Haute (IN) Volunteer Fire Department died after suffering a heart attack at his station. Barter had just returned from an emergency medical call in very hot weather.

Incident 39

On June 24, Lt. Stephen Minehan of the Boston (MA) Fire Department died after leading his company in a successful search for two other trapped firefighters at a blaze in a vacant waterfront warehouse. Minehan apparently became disoriented in the heavy smoke conditions and was separated from his company as they rescued the trapped firefighters. He radioed that he was trapped but several rescue efforts to find him were unsuccessful. He died of smoke inhalation and his body was recovered by his company several hours later.

Incident 40

On June 28, Chief Clifford Harris of the Rusk (TX) Volunteer Fire Department responded to a house fire. After the fire was knocked down, Chief Harris entered the structure to assist with overhaul operations in the area of origin. Chief Harris was not wearing turnout gear or SCBA. After several minutes, Harris left the structure and collapsed in cardiac arrest. He had heart bypass surgery 10 years earlier. The fire had burned several hundred plastic video cassette tapes which gave off toxic gases. Harris death was attributed to a heart attack caused by inhalation of toxic gas; no autopsy was performed.

Incident 41

On June 29, Fire Marshal Anthony Bullard of the Greenville (TX) Fire Department suffered a cerebral hemorrhage during physical education class at the police academy. He died on June 30.

Incident 42

On July 6, fourteen wildland firefighters lost their lives when a wind shift resulted in a blow-up fire condition that trapped them on the uphill and downwind position from the fire on Storm King Mountain, Colorado. The fourteen firefighters included smokejumpers Don Mackey, Roger Roth, and James Thrash; Prineville Hot Shots Jon Kelso, Kathi Beck, Scott Blecha, Levi Brinkley, Bonnie Holtby, Rob Johnson, Tami Bickett, Doug Dunbar, and Terri Hagen; and helitack crew members Richard Tyler and Robert Browning. Browning and Tyler were killed when their escape route was cut off by a large drop and they were overrun by the fire. The other firefighters were killed as they moved towards the ridgeline to escape the fire advancing towards them from below. According to witness accounts, the firefighters were unable to see how dangerous their position had become because of a small ridge below them. They had been moving slowly and were still carrying their equipment as the fire blew up behind them to a height of over 100 feet. At this point the crew dropped their tools and made an uphill dash for the top of the mountain but only one person made it over to survive. The fire overran the remaining twelve firefighters and reportedly reached a height of 200 to 300 feet as it crossed over the ridge. It was estimated to be moving at between 10 and 20 miles per hour at the time of the blow-up.

Several other firefighters in various other locations on the mountain became trapped by the flames but were able to make it to safe positions or deploy their emergency shelters. Post incident investigations have determined that the crews fighting the fire violated many safety procedures and standard firefighting orders. The weather conditions prevalent that day had forecast a "red flag" Learning, the most dangerous wildfire conditions.

Incident 43

On July 12, Pilot Robert Boomer of Briles Wings and Helicopter and Helitack Firefighters Sean Gutierrez and Sam Smith of the U.S. Forest Service were killed when their helicopter crashed while transporting them for an initial attack on. the Guide Fire burning in the Black Range of the Gila National Forest. Two other crew members were injured in the crash.

Incident 44

On July 23, Firefighter Michael Shaughnessy of the Cleveland Fire Department was killed when he fell off the roof of his fire station.

Incident; 45

On July 27, Firefighter Paul Hodges died after suffering a heart attack while driving a tanker truck on a wildland fire. Hodges was a contract employee for the USFS and a volunteer with the Chelan County Fire Protection District #9.

Incident 46

On July 29, Pilots Bob Kelly and Randy Lynn of Neptune, Inc., were killed when their airtanker crashed after dropping retardant on a wildfire.

Incident 47

On August 3, Sergeant John Nutter of the Louisville (KY) Division of Fire was killed when the roof collapsed under him while performing ventilation at a fire in a storage facility. Sgt. Nutter fell into a storage area where, according to investigators, he was able to force his way into a hallway but was then trapped by interlocking doors and heavy fire conditions. Rescue efforts were hampered by maze-like conditions in the building. Sgt. Nutter was found with a depleted air bottle and dislodged facepiece. He had been exposed to fire conditions that exceeded the protective envelope provided by his turnout gear. Efforts to revive him were unsuccessful. He was wearing an operating PASS device but it was in the off position. He died of smoke inhalation and burns.

Incident 48

On August 7, Captain Wayne Smith of the New York City Fire Department was critically injured while conducting search and rescue operations on an upper floor of a building when he was trapped by high heat and heavy smoke conditions. Captain Smith was burned over 40 percent of his body and received severe smoke inhalation injuries to his lungs. He died on October 4 from his injuries. Fourteen other firefighters were injured in the blaze. Initial operations were hampered by a faulty fire hydrant across the street from the building.

Incident 49

On August 13, an air tanker crashed enroute to a wildfire in Kern County, California. Robert Buc, Joe Johnson, and Shawn Zaremba, the flight crew from the Hemet Valley Flying Service, were killed in the crash.

Incident 50

On August 13, Firefighter James Harvey of the Greenwood (IN) Fire Department suffered a fatal heart attack during training.

Incident 51

On August 18, Firefighter Herbert Smith of the Shelby (AL) Volunteer Fire Department suffered a fatal heart attack at a fire.

Incident 52

On August 18, Firefighter Sam McCarty of the Harding County Fire District #2 suffered a heart attack and died while cutting a fire line with a road grater at a grass fire.

Incident 53

On August 8, Firefighter David Castro of the U.S. Forest Service died of traumatic injuries suffered when his water tanker truck overturned on the Quincy-Oroville Highway while enroute a wildfire.

Incident 54

On August 8, Sergeant Craig Drury of the Highview (KY) Fire District was caught in a flashover while making entry into a single story house. Sgt.

Drury suffered severe burns to his lungs that eventually caused his death. The fire started by an arsonist who started the fire because of an interracial adoption.

Incident 55

On August 25, Firefighter Sydney Bruce Maplesden, Jr., of the Oregon Department of Forestry died when he was overrun by a wildfire while attempting to cut a fire break with a bulldozer.

Incident 56

On August 27, Firefighter Paul MacMurray of the Hudson Falls (NY) Volunteer Fire Department responded as part of an engine company to a fire on the first floor of in a three story hotel. Assigned to search for and rescue occupants on the second floor, MacMurray and another firefighter successfully evacuated several victims while attempts to extinguish the fire were initiated below them. Upon their return to continue the search, conditions quickly changed from a light haze of smoke to black smoke with high heat. conditions. MacMurray and his partner became separated in their attempt to locate the stairwell and get out of the building. The other firefighter made several efforts to locate MacMurray, but was forced to retreat due to untenable conditions. Several rescue efforts were made but heavy fire conditions eventually forced the evacuation of all fire personnel to defensive positions as the entire structure burned. MacMurray's body was recovered the following day. The fire was incendiary in nature.

Incident 5 7

On August 27, Assistant Chief Gerald Murray of the Windham (NY) Hose Company #1 died when his personal vehicle crashed while responding to an unknown type of fire.

Incident 58

On September 3, Captain Earl Detty of the Union Township (OH) Fire Department responded to a report of a fire in the woods. Arriving, on the scene, Captain Detty discovered a legal campfire. He suffered a heart attack and died while walking back to his fire engine.

Incident 59

On September 6, Firefighter Recruit Dwight Smith of the Memphis (TN) Fire Department collapsed in cardiac arrest after a jog during a physical training session. Firefighter Smith was in his fourth week of firefighter recruit class.

Incident 60

On September 11, Lieutenant Dewey Henry of the Metro-Dade (FL) Fire and Rescue Department was killed after being trapped under rolls of carpet and debris when the roof collapsed during a fire in a carpet warehouse.

Incident 6 1

On September 13, Chief Gus Fullbright of the Sallisaw (OK) Fire Department assumed incident command of a fire in a fully involved unattached garage with a vehicle inside. Firefighters had pulled a 1" booster line to protect the house exposure and were preparing to deploy larger attack lines in a defensive mode when a loud hiss was heard and an explosion took place. Seven firefighters were burned, including Chief Fullbright who was standing 40-50 feet away wearing only his helmet. Chief Fullbright died of his burn injuries two weeks after the incident.

Incident 62

On September 15, Robert L. Johnson, a driver for the Bureau of Land Management at the National Interagency Fire Center, was killed when his truck was struck by another vehicle while enroute with supplies to a wildfire base camp.

Incident 63

On September 22, Captain James Certain of the Scenic Loop (TS) Volunteer Fire Department was killed when the 1972 Chevrolet 3000 gallon tanker truck he was driving overturned at an intersection while enroute to a house fire. Captain Certain died of his injuries at the scene.

Incident 64

On September 23, John King, a civilian serving as a flight engineer for a US Army Reserve CH-47D helicopter, died of traumatic injuries when he was struck by a helicopter rotor blade during crash while attempting to pick up firefighters during the Chicken Complex fire in McCall, Idaho. King's unit had been assigned to provide emergency support to the United States Forest Service for wildland fire suppression.

Incident 65

In September, James Harris of the Mechanicville (NY) Fire Department suffered a fatal heart attack after running to the fire station to respond on a fire call.

Incident 66

On October 6, Fire Management Officer Daren Smith of the U.S. Forest Service was killed when he was struck by a falling tree while clearing a fire road.

Incident 67

On October 13, Firefighter Roy Stephenson of the #1 Green Township (IN) Volunteer Fire Department died after suffering a heart attack while operating a pumper at a working structure fire.

Incident 68

On October 29, Firefighter Michael DeLane of the Newark (NJ) Fire Department was climbing down an aerial ladder after roof operations a two alarm fire. As DeLane passed a saw to a fellow firefighter, it came in close proximity to a power line, electrocuting both of them. DeLane was killed and the other firefighter injured. One civilian died in the fire, which had been extinguished at the time of the accident.

Incident 69

In October, Elias Ovsiovitch of the Hillcrest (NY) Fire Company #1 suffered a fatal heart attack while performing clerical duties at the fire station.

Incident 70

On November 8, Firefighter Brian D. Sutton, Sr., of the Enterprise (NJ) Fire Company was stricken by a heart attack as he was hooking up to a hydrant during pump operations at a house fire.

Incident 71

On November 9, Firefighter Richard Liddy of the Basking Ridge (NJ) Fire Company #1 suffered a fatal heart attack while pulling hose at the scene of a house fire.

Incident 72

On November 12, Firefighter Edward Freeman of the Memphis (TN) Fire Department suffered a fatal heart attack after returning to his station after an auto fire.

Incident 73

On November 20, Firefighter Mary Jo Brown of the U.S. Forest Service died of smoke inhalation after her position was overrun by a rapidly moving wildfire. Two other firefighters deployed their personal shelters and survived the fire.

Incident 74

On November 23, Fire Management Officer Roger Evans of the U.S. Forest Service was killed when a 106mm rifle exploded during training for avalanche control, part of his collateral duties.

Incident 75

On November 28, Firefighter Dale Nclboeck of the Mayfiled Heights (OH) Fire Department collapsed and died of an apparent heart attack-as he was helping carry a patient to an EMS unit during a rescue call.

Incident 76

On December 6, Firefighter Dwight Burger of the South Danesville (NY) Volunteer Fire Department died in an apparatus crash enroute to an emergency.

Incident 77

On December 10, Captain Jesse Shockley, Jr. of the Fort Bragg (NC) Fire Department died after suffering a heart attack during a training session on ladders.

Incident 78

On December 24, Firefighter Lionel Hoffer of the Milwaukee Fire Department died at a fire in a church when he fell through a hole in the second floor. Hoffer had been operating an attack line in the church with an engine company when he went to check a room for fire extension. His crew heard him call for help and found him hanging from a hole in the floor, but they were unable to keep him from falling. He fell approximately 12 feet into the first floor. Rescue crews tried to reach him, alerted by his PASS device which had activated, but were hampered by the collapsed floor in the front of the building and barred security doors at the rear of the building. Rescue crews eventually fought their way through heavy heat and smoke conditions to Hoffer's location, removed him from the building, and administered emergency care. His air supply had been exhausted and his death has been attributed to smoke inhalation.

Incident 79

On December 26, Firefighter Evan Buchholtz of the Poy-Sippi (WI) Fire Department suffered a fatal heart attack while performing duties at a house fire.

Incident 80

On December 27, Chief Engineer Steven Colona of the Melfa (VA) Volunteer Fire Department died of traumatic injuries when the tanker truck he was driving overturned on the way to reported fire in a. chicken house. Another firefighter was severely injured in the accident. The call turned out to be a false alarm.

Incident 81

On December 27, Firefighter Thomas Wylie suffered carbon monoxide poisoning in a structure fire. He died in 1995.

www.ingramcontent.com/pod-product-compliance
Lightning Source LLC
Chambersburg PA
CBHW081222170526
45165CB00009B/2908